ISO 45001 Documentation Kit

For Small Enterprises

Occupational Health and Safety: Practical Templates and Tools for Implementation, Compliance, and Audit Readiness in All Business Sectors

Dr. Mohamed Ali Ibrahim

ISO 45001 Documentation Kit for Small Enterprises

Occupational Health and Safety: Practical Templates and Tools for Implementation, Compliance, and Audit Readiness in All Business Sectors

Dr. Mohamed Ali Ibrahim

ISO 45001 Documentation Kit for Small Enterprises

This documentation kit is dedicated to small business owners and their teams who work every day to maintain safe and healthy workplaces. Your commitment to protecting your Workers while managing the challenges of daily operations is commendable. This kit is designed to support your efforts with practical tools that help strengthen compliance, improve control, and build a safer working environment.

DR. MOHAMED ALI IBRAHIM

Dr. Mohamed Ali Ibrahim is a visionary entrepreneur and expert in ISO 45001 and occupational health and safety management. with over 25 years of experience in market expansion, operational excellence, and business innovation. As the founder of IQC-Vienna, he helps organizations achieve sustainable growth, improve their Quality Management Systems (QMS), and drive profitability through strategic quality-focused solutions. His expertise is reflected in his book, **Mastering Contract Management: A Practical Guide to Effective Administration and Dispute Resolution**, where he shares his vast knowledge of contract management, risk mitigation, and dispute resolution to empower professionals in managing successful, compliant contracts.

Visionary Leader in Quality Management and International Standards

He is the author of over 80 professional publications in **English**, **Arabic**, and **German**, focusing on **business management**, **quality and risk management**, and **translation studies**. His latest work is ***"Introduction to ISO 45001:2018."***

As an MBA lecturer, expert in **digital transformation**, and **Senior Lead Auditor**, Dr. Ibrahim specializes in international standards, including:

- **ISO 27001**
- **ISO 22301**
- **ISO 45001**
- **ISO 9001**
- **ISO 14001**
- **ISO 18587**
- **ISO 20771**

He is regarded as a dedicated advocate for **institutional quality** and works towards **resilient** and **future-proof business structures**.

His expertise is valued across industries, with clients ranging from **ministries** and **government agencies** to **public figures** and leading organizations from various sectors.

With a unique combination of **academic depth** and **practical implementation skills**, Dr. Ibrahim is seen as a beacon of excellence in promoting **high-quality standards** and **innovative practical solutions**. His significant contributions to **quality management** and **international standardization** have also made him a sought-after **strategic advisor**.

In particular, his ability to develop **complex strategies** and **operational procedures** for ministries, authorities, and large corporations has made him the preferred consultant for demanding clients worldwide.

Dr. Ibrahim's consulting work is characterized by a deep understanding of the multifaceted **challenges** and **opportunities** in various sectors, always guided by **precision**, **innovation**, and **strategic foresight**.

His career stands for a sustainable commitment to raising **industry standards** and promoting **quality**, **excellence**, and **visionary thinking**. His influence extends far beyond operational issues and inspires all those striving to be **leaders** in their industries and act with a **future-oriented mindset**.

Professional and academic achievements:

Dr. Mohamed Ali Ibrahim specializes in **multilingual communication** and has a solid academic background in **English**, **Arabic**, and **German**. His Master's degree at **Karl-Franzens University Graz** (Austria) focused on **economic**, **legal**, and **strategic technical translation**, fostering **precise, culturally sensitive interpretation** across various contexts. He holds a **PhD** in **Quality and Risk Management Systems**, focusing on the development and implementation of concepts for **quality assurance** and **risk minimization** to achieve **organizational excellence** and ensure compliance with **international standards**.

In addition, he holds **further diplomas** in **business management**, **strategy**, and **quality management**, awarded by the **Austrian Standards Institute (ASI)** and the **British Standards Institution (BSI)**. These qualifications underscore his comprehensive expertise in **strategic leadership**, **operational excellence**, and **compliant quality assurance**.

ISO Standard	Area of Specialization
ISO 22301	Business Continuity Management
ISO 9001	Quality Management Systems
ISO 45001	Occupational Health and Safety Management Systems
ISO 14001	Environmental Management Systems
ISO 18587	Machine Translation Post-Editing
ISO 20228	Interpreter Services - Legal Interpreting
ISO 13611	Community Interpreting
ISO 18841	Interpreter Services
ISO 21001	Educational Management Systems
ISO 17100	Translation Service Providers
ISO 21998	Healthcare Interpreting
ISO 23155	Conference Interpreting
ISO 20771	Legal Translation
ISO 27001	Information Security Management Systems

Authorship and Contributions

Publications

Dr. Mohamed Ali Ibrahim is the author of over 80 professional books published on Amazon. His key areas of focus include **quality and risk management**, **business leadership**, **communication strategies**, and **translation studies**.

Professional Expertise

He is internationally recognized for his work in **consulting**, **training**, and **certification**. With over 750 quality audits conducted worldwide, Dr. Ibrahim has significantly contributed to strengthening global standards in the fields of **translation** and **interpreting**.

Dr. Ibrahim undertook strategic consulting assignments and led specialized workshops in **quality management** for high-ranking government institutions, including the **Ministry of Culture**, the **Royal Court**, and the **Saudi Industrial**

Development Fund (SIDF) in Saudi Arabia. The focus was always on optimizing quality standards in **culture**, **administration**, and **industry**. He played a pivotal role in the introduction and implementation of **ISO certifications**, particularly **ISO 17100** (Translation Services) and **ISO 9001** (Quality Management), and supported organizations in building efficient translation and communication departments.

International Activities

Dr. Ibrahim is currently engaged in consulting and training activities in multiple countries, including Austria, the USA, the United Kingdom, **Egypt**, the **United Arab Emirates**, **Saudi Arabia**, **Algeria**, and **Jordan**. This international presence underscores his strong commitment to the global development of **quality standards** and **professional language services**.

Region/Institution	Activity/Role
Kingdom of Saudi Arabia	Strategic consulting and leading workshops in **Quality Management** for high-ranking government institutions (e.g., **Ministry of Culture, Royal Court, SIDF**)
ISO Certifications	Led the introduction and implementation of **ISO 17100** (Translation Services) and **ISO 9001** (Quality Management)
International	Engaged in consulting and training in **Austria, USA, UK, Egypt, UAE, Saudi Arabia, Algeria, Jordan**

Lectures & Teaching

Since 1998, Dr. Mohamed Ali Ibrahim has been a sought-after keynote speaker at international conferences and a lecturer at **Vienna University of Applied Sciences**. In his lectures, seminars, and courses, he imparts deep knowledge and practical best practices in **quality management**, **translation studies**, and **risk management**.

Awards:

- **Top Expert 2025** by **ERFOLG-Magazine** – Recognized as one of the leading consultants in **quality and risk management** in the **DACH** region.
- **Austrian State Appreciation Award** – For demonstrable business growth and outstanding operational achievements.
- **Recognition for Outstanding Contributions** under German and British award programs.

- **Expert of the Year 2021–2024** – For groundbreaking, customer-oriented solutions and innovative market strategies.
- **Reviewer at the Saudi Center for Accreditation**.
- **International EFQM Assessor** (European Foundation for Quality Management).
- **Recipient of the International German Prize**.
- **Awarded the Best Practice Award Austria**.
- **Winner of the SABRE Award** (United Kingdom) – also known as the "PR-Oscar" for excellent international communication achievements.

Lifetime Commitment

Dr. Ibrahim's works, published on **Amazon**, reflect his lifelong mission: to establish, develop, and pass on **quality management** as a central tool for excellence in **business**, **education**, and **society**.

Publications

Dr. Mohamed Ali Ibrahim has authored over 80 books on Amazon. These publications cover topics in **business management**, **quality management**, and **real estate management**.

Explore Dr. Ibrahim's Books on Amazon

https://www.amazon.com/stores/author/B09YXDBTV5

LinkedIn Profile

https://www.linkedin.com/in/ph-dr-mohamed-ali-ibrahim/

Academy IQC Vienna

Welcome to Academy IQC Vienna — your hub for innovative online learning and resources to boost occupational health and safety management, ISO 45001:2018 compliance., and business performance through tailored courses, e-forms, and coaching.

Service	Description	Link
E-Workshop	Join interactive e-workshops that provide in-depth training on **ISO standards**, **quality management**, and **risk management**. Engage in practical exercises.	https://academy.iqc-vienna.com/e-workshop/
E-Coaching	Get personalized coaching from industry experts, whether you need guidance on ISO certification, audit preparation, or process optimization.	https://academy.iqc-vienna.com/e-coaching/
Online Courses	Take advantage of our self-paced online courses designed to help you master **ISO standards**, enhance **quality management**, and develop necessary skills.	https://academy.iqc-vienna.com/online-courses/
Documentation Kits	Streamline your ISO 45001 implementation with comprehensive documentation kits., including templates for **policies**, **procedures**, and **forms**.	https://academy.iqc-vienna.com/documentation-kits/

Introduction

An effective occupational health and safety management system depends not only on good intentions, but also on clear, practical documentation. Small enterprises often face real challenges in maintaining organized records, consistent procedures, and usable templates that support daily OH&S activities. For that reason, this documentation kit has been developed as a practical companion to ISO 45001:2018, helping small enterprises manage essential documentation in a structured, simple, and professional way.

This kit is designed to support implementation, compliance, and audit readiness by providing practical templates and tools that can be adapted to real workplace needs. Rather than relying on scattered records or inconsistent formats, organizations can use this kit to strengthen control over key OH&S activities such as policy management, based on based on risk assessment methods defined by the organization in line with ISO 45001:2018 Clause 6.1.2" to align with risk-based thinking, risk assessment, training, corrective action, and audit preparation. The purpose is to make documentation more usable, more consistent, and more effective in supporting a functioning OH&S management system.

Benefits of This Documentation Kit

This documentation kit offers practical value by helping small enterprises create and maintain documentation that supports the daily operation of their OH&S management system. Clear templates reduce confusion, save time, improve consistency, and help ensure that important OH&S activities are recorded in a controlled and understandable way. This is especially important in small businesses, where limited resources and overlapping responsibilities can make documentation difficult to manage without a structured system.

Another key benefit is that the kit supports better accountability and follow-up. When forms, logs, registers, and checklists are properly structured, it becomes easier to assign responsibilities, track actions, review results, and retain evidence of what has been done. The kit also supports stronger audit readiness by helping organizations maintain documentation that is relevant, organized, and aligned with ISO 45001:2018 requirements and practical OH&S management needs.

Why Choose This Documentation Kit

This documentation kit should be chosen because it is designed specifically for small enterprises that need practical, professional, and usable OH&S documentation. It is not intended to create unnecessary paperwork. Instead, it provides focused templates and tools that help organizations manage documentation more effectively while keeping their OH&S system clear and workable.

Another reason to choose this kit is its practical alignment with ISO 45001:2018 terminology and structure. The templates are intended to support real management activities such as planning, operational control, competence management, nonconformity handling, and audit readiness. This makes the kit useful not only for documentation purposes, but also for improving consistency, control, and confidence across the OH&S management system.

Who This Documentation Kit Is For

This documentation kit is intended for small business owners, managers, supervisors, OH&S coordinators, consultants, internal auditors, and others who are responsible for maintaining occupational health and safety documentation in small enterprises. It is especially useful for organizations that need practical templates to support implementation, compliance, and preparation for internal or external audit activities.

It is also suitable for businesses that want to improve the quality, consistency, and usability of their OH&S records without creating unnecessary complexity. Whether the organization is establishing documentation for the first time, improving an existing system, or preparing evidence for review, this kit provides structured tools that can support more effective OH&S management in daily business operations.

Table of Contents

PART 1

Understanding Certification and the Audit Process for Small Enterprises

Setting the foundation for achieving ISO 45001:2018 certification and preparing for the audit process.

CHAPTER 1

Introduction to ISO 45001 Certification for Small Enterprises

The Importance of ISO 45001 for Small Businesses

ISO 45001:2018 is the international standard for Occupational Health and Safety (OH&S) management systems. It provides a framework for organizations to manage and improve employee safety, reduce workplace risks, and create a healthier work environment. While this standard applies to organizations of all sizes, its adoption in small enterprises offers a unique set of advantages. This chapter will delve into the significance of ISO 45001 for small businesses, particularly focusing on how it can foster growth, improve workplace culture, and safeguard both employees and the business.

What Is ISO 45001:2018?

ISO 45001:2018 is a global standard for creating an OH&S management system that enables businesses to proactively reduce risks and enhance the safety and well-being of their employees. This certification helps businesses assess, monitor, and mitigate risks associated with occupational health and safety by setting a structured approach for risk management, training, and incident reporting.

For small businesses, the implementation of ISO 45001 means creating a system that not only complies with legal requirements but also goes beyond to promote a culture of continuous improvement in health and safety practices.

The Value of ISO 45001 for Small Businesses

ISO 45001 provides numerous benefits to small businesses, whether they are a growing enterprise or a well-established local company. Here's why ISO 45001 is so crucial:

1. Legal Compliance and Risk Management

Stage 1 assesses documentation and system implementation readiness; it may identify potential gaps but does not replace Stage 2 effectiveness evaluation. such as health and safety laws, industry standards, and local regulations. The risk of workplace accidents or health hazards is managed proactively, reducing the chances of costly legal actions, fines, and penalties.

- **Regulatory Adherence**: The standard helps small businesses stay on top of ever-changing safety regulations and avoid legal non-compliance.
- **Legal Safety**: With ISO 45001, small enterprises can confidently show they're taking steps to meet national and international safety standards.

2. Enhanced Workplace Safety and Reduced Incidents

Adopting ISO 45001 leads to a direct reduction in workplace injuries, illnesses, and accidents. Small businesses are often more vulnerable to operational disruptions due to a lack of formalized health and safety practices. ISO 45001 helps identify, control, and mitigate risks, creating safer working conditions.

- **Injury Reduction**: Small businesses with ISO 45001 certification have demonstrated significantly lower injury rates.
- **Proactive Safety Culture**: The standard fosters a proactive approach to safety, where hazards are anticipated, and preventive measures are implemented before accidents occur.

3. Increased Employee Morale and Engagement

When employees feel safe and valued, their morale improves, which in turn boosts productivity and reduces absenteeism. ISO 45001 emphasizes employee participation in health and safety programs, which can foster a positive organizational culture.

- **Employee Well-being**: The certification focuses on the physical and mental health of employees, creating a supportive work environment.
- **Trust and Loyalty**: Employees are more likely to remain with businesses that prioritize their safety and health, resulting in better retention rates.

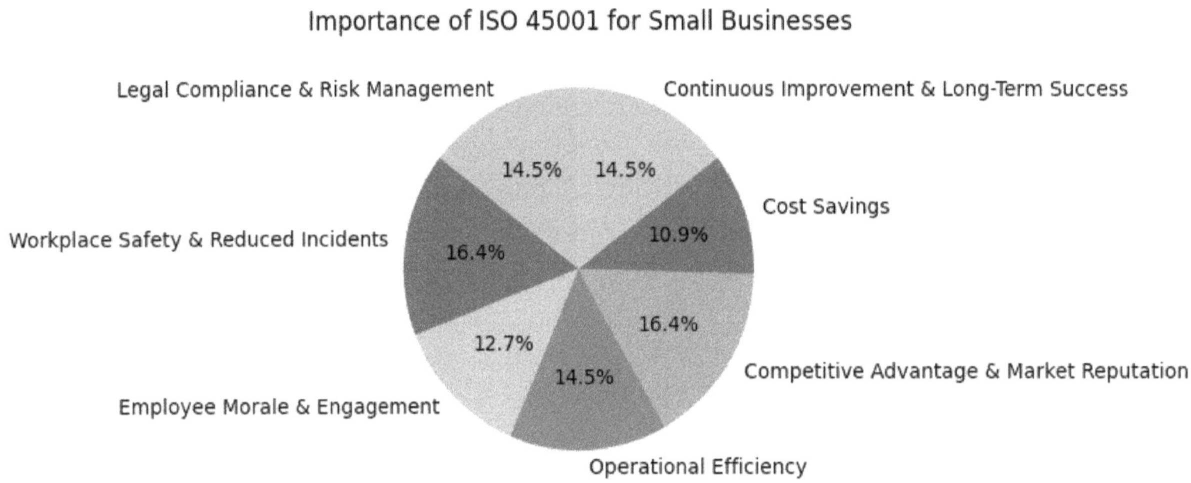

4. Improved Operational Efficiency

ISO 45001 provides a structured framework for small businesses to improve internal processes related to health and safety. By standardizing risk management procedures, small businesses can streamline their operations, reduce waste, and enhance resource utilization.

- **Resource Optimization**: A well-implemented OH&S system means fewer disruptions, less downtime, and a more focused workforce.
- **Increased Productivity**: With fewer incidents and better risk management, businesses can maintain smoother operations, leading to increased efficiency and productivity.

5. Competitive Advantage and Market Reputation

Having ISO 45001 certification can set small businesses apart from competitors. Customers, suppliers, and stakeholders are increasingly prioritizing organizations that uphold high safety standards, and ISO 45001 is an internationally recognized mark of credibility.

- **Customer Confidence**: ISO 45001 certification builds trust with customers, as they can rely on your company's commitment to health and safety.
- **Business Growth**: Many tenders and contracts, especially in sectors like construction, manufacturing, and healthcare, require businesses to be ISO 45001 certified, giving those that are certified a competitive edge.

6. Cost Savings

While implementing an OH&S system requires an initial investment, the long-term savings from avoiding accidents, reducing insurance premiums, and increasing productivity can be substantial.

- **Insurance Premium Reduction**: Insurance companies often offer lower premiums to businesses with ISO 45001 certification because it demonstrates a commitment to managing risk.
- **Fewer Absences**: Reduced injury and illness rates result in fewer employee absences and related costs, such as compensation claims.

7. Continuous Improvement and Long-Term Success

ISO 45001 emphasizes continuous improvement. Small businesses are not just compliant; they are continually enhancing their OH&S performance. This approach helps businesses stay adaptable and resilient, ensuring they evolve with industry trends, employee needs, and regulatory changes.

- **Adaptation and Growth**: The standard ensures that businesses are always improving their systems, keeping up with both safety regulations and best practices.
- **Sustainable Practices**: ISO 45001 leads small enterprises toward sustainable business practices that benefit both the organization and society.

Conclusion: Why ISO 45001 Is Critical for Small Businesses

For small enterprises, ISO 45001 offers a framework for creating a safe, healthy, and compliant workplace. It goes beyond just meeting the minimum legal requirements and creates a culture of safety, health, and continuous improvement. Implementing ISO 45001 not only protects the business and its employees but also provides a competitive advantage in the marketplace, ensuring long-term success. By choosing to adopt ISO 45001, small businesses can proactively address health and safety risks while contributing to a more secure and productive work environment.

Overview of the Certification Process for Small Enterprises

The certification process for **ISO 45001:2018** in small enterprises involves several key steps designed to ensure that businesses are compliant with the standard's requirements for Occupational Health and Safety (OH&S) management systems. While the process may seem complex, breaking it down into manageable phases helps small enterprises implement effective systems for improving workplace safety and protecting employee health.

This process ensures that small enterprises create, maintain, and continuously improve OH&S practices, mitigating risks and ensuring regulatory compliance. Below is a detailed explanation of each phase of the certification process, followed by a table summarizing the key steps.

Step-by-Step Breakdown of the Certification Process

1. Understanding ISO 45001:2018 Requirements

Before initiating the certification process, it's crucial for small businesses to gain a clear understanding of what ISO 45001:2018 entails. This standard specifies the requirements for an OH&S management system, including:

- **Based on based on risk assessment methods defined by the organization in line with ISO 45001:2018 Clause 6.1.2" to align with risk-based thinking. and risk assessment**: Identifying potential safety hazards and assessing the risks they pose.

- **Legal compliance**: Ensuring adherence to relevant occupational health and safety laws.
- **Employee involvement**: Ensuring that employees are engaged in safety management and are aware of their responsibilities.
- **Management responsibility**: Senior management must demonstrate leadership and commitment to OH&S.

2. Pre-Certification Assessment

Small businesses should begin by conducting a pre-certification assessment or gap analysis. This step helps determine the current state of the organization's OH&S management system in comparison to the ISO 45001 requirements. The pre-assessment should focus on:

- **Identifying any gaps** between current safety practices and ISO 45001 standards.
- **Documenting existing procedures**, policies, and safety management systems.
- **Determining resource requirements** for achieving certification.

3. Developing the OH&S Management System

Once gaps are identified, small businesses should start developing their OH&S management system (OHSMS). This includes:

- **Establishing a clear OH&S policy**: A documented policy that outlines the company's commitment to workplace safety and health.
- **Risk assessments and controls**: Performing a thorough assessment of potential hazards and implementing control measures.
- **Employee training and awareness**: Ensuring employees are trained on safety protocols and their roles in maintaining a safe working environment.
- **Documenting processes**: Developing and documenting standard operating procedures (SOPs) related to safety measures, emergency responses, and risk management.

4. Implementing the OH&S Management System

Once the system is developed, businesses should implement it throughout their operations. This phase focuses on:

- **Communicating the system**: Ensuring that all employees are aware of the new policies, procedures, and safety measures.
- **Monitoring compliance**: Regularly checking to ensure that the system is being followed and is effective in reducing risks.
- **Collecting performance data**: Gathering data on workplace incidents, near misses, and overall safety performance.

5. Internal Audit and Review

Before seeking external certification, small businesses should conduct an internal audit to ensure the OH&S system is fully functional and compliant with ISO 45001. The audit will help identify:

- **Non-conformities**: Issues that may prevent the system from meeting ISO 45001 standards.
- **Areas for improvement**: Opportunities to strengthen safety practices and reduce risks further.
- **Management review**: Senior management should review audit results to ensure the system meets business goals and regulatory requirements.

6. Certification Audit

Once internal audits and reviews are complete, small businesses can request an external audit from an accredited certification body. The audit will assess:

- **Compliance with ISO 45001 standards**: The certification body will evaluate whether the OH&S system aligns with ISO 45001:2018.
- **Effectiveness of the system**: The auditors will check if the OH&S system has effectively reduced risks and improved employee safety.
- **Corrective actions**: If non-conformities are found, businesses will need to take corrective actions before certification is granted.

7. Ongoing Maintenance and Surveillance

After achieving certification, small businesses must maintain their ISO 45001 compliance through regular monitoring and surveillance audits. The certification body will typically conduct annual surveillance audits to ensure continued compliance. Businesses should:

- **Monitor and review** their OH&S management system regularly.
- **Make improvements** based on feedback from audits and performance reviews.
- **Engage employees** to ensure they remain active in safety practices.

Table: Key Steps in the ISO 45001 Certification Process for Small Enterprises

Step	Action	Details
1. Understand ISO 45001	Study the standard's requirements	Familiarize with based on based on risk assessment methods defined by the organization in line with ISO 45001:2018 Clause 6.1.2" to align with risk-based thinking., legal compliance, and the roles of management and employees
2. Pre-Certification Assessment	Conduct a gap analysis	Identify current gaps between your OH&S system and ISO 45001 standards
3. Develop OH&S System	Create an OH&S management system	Establish policies, risk assessments, procedures, and training programs
4. Implement OH&S System	Put the OH&S system into practice	Ensure employee engagement, communicate policies, and monitor compliance
5. Internal Audit and Review	Perform an internal audit and review	Identify non-conformities and areas for improvement before external audit
6. Certification Audit	Request an external audit from a certification body	The audit evaluates compliance with ISO 45001 standards and effectiveness of the OH&S system
7. Ongoing Maintenance	Conduct regular monitoring and surveillance audits	Continue improving the system and undergo annual surveillance audits

Conclusion

The certification process for ISO 45001:2018 may seem complex, but it is highly beneficial for small enterprises. By following these steps and ensuring commitment to continuous improvement, businesses can enhance workplace safety, reduce risks, and meet legal obligations. Moreover, the process itself promotes a culture of safety and wellbeing that benefits both employees and the organization's long-term success.

Benefits of Certification for Small Enterprises and Businesses

Achieving **ISO 45001:2018** certification offers numerous advantages for small businesses. Certification provides a formalized, internationally recognized system to improve Occupational Health and Safety (OH&S) practices within a company. The benefits of ISO 45001 for small enterprises go beyond compliance with legal obligations—they encompass improved safety performance, cost savings, employee morale, and business growth. Below is a detailed explanation of the key benefits for small businesses:

1. Improved Health and Safety Standards

ISO 45001 ensures that small businesses adhere to the highest standards of occupational health and safety. By implementing the standard, businesses can identify hazards, assess risks, and establish measures to eliminate or mitigate those risks. This proactive approach leads to:

- **Fewer workplace injuries and illnesses**: A key outcome of implementing an OH&S management system is a reduction in workplace accidents and health-related issues. By addressing safety risks before they lead to incidents, businesses create a safer environment for employees.
- **Preventative measures**: ISO 45001 focuses on preventing hazards rather than just responding to incidents. This reduces the likelihood of workplace accidents, illnesses, and near-miss events.
- **Better incident reporting**: The certification encourages employees to report potential hazards, contributing to an overall safer work environment.

2. Legal Compliance and Risk Management

Compliance with legal health and safety regulations is mandatory for businesses of all sizes. However, small businesses, due to their limited resources, might struggle to stay updated with ever-changing regulations. ISO 45001 helps small enterprises:

- **Ensure adherence to OH&S laws**: ISO 45001 ensures that businesses comply with local, national, and international health and safety laws. This helps businesses avoid fines and legal liabilities related to non-compliance.
- **Risk management and mitigation**: The standard requires businesses to perform risk assessments, which helps identify potential hazards, assess their severity, and implement effective controls to reduce the likelihood of accidents and incidents.
- **Liability reduction**: By proactively managing health and safety risks, small businesses reduce their liability for accidents, ensuring they avoid costly lawsuits or regulatory penalties.

3. Cost Savings

Though achieving ISO 45001 certification requires an investment of time, money, and resources, the long-term financial benefits far outweigh these initial costs. The certification helps small businesses:

- **Lower insurance premiums**: Many insurance companies offer reduced premiums to businesses that demonstrate commitment to safety by being ISO 45001 certified. This is because these businesses are less likely to experience accidents, which can result in fewer insurance claims.
- **Reduce compensation claims**: A safer workplace means fewer injuries and illnesses, which leads to fewer Workers' compensation claims. This reduces operational costs and minimizes the financial burden associated with workplace accidents.
- **Improved productivity and efficiency**: A safe workplace leads to less downtime and fewer disruptions. Employees are healthier, more engaged, and more productive, which enhances the overall efficiency of the business.

4. Employee Engagement, Well-being, and Morale

Employees are a critical asset to any business, and their safety and well-being directly affect the success of an organization. ISO 45001 certification improves employee engagement by:

- **Boosting employee morale**: Employees working in a safe and healthy environment are more likely to feel valued and appreciated, which improves their overall morale. A positive workplace culture enhances job satisfaction, which can lead to higher retention rates.
- **Increased employee involvement**: ISO 45001 emphasizes the involvement of employees in health and safety matters. Employees are encouraged to participate in safety committees, report hazards, and contribute ideas to improve safety practices. This increases engagement and fosters a sense of responsibility and ownership.
- **Better health outcomes**: With ISO 45001, small businesses prioritize both physical and mental health. Reducing workplace stressors, addressing potential hazards, and offering wellness programs contribute to better overall employee health, leading to fewer sick days and improved productivity.

5. Enhanced Business Reputation and Competitive Advantage

ISO 45001 certification gives small businesses a competitive edge by demonstrating their commitment to health and safety. Certification:

- **Builds trust with customers and partners**: Clients, suppliers, and business partners are more likely to trust a business that is certified ISO 45001. It assures them that the business is committed to providing a safe and healthy environment for employees and customers.
- **Improves marketability**: Certification allows small businesses to differentiate themselves in the market. As safety becomes an increasing concern for customers, companies with ISO 45001 certification stand out as responsible and trustworthy.
- **Opens new opportunities**: Many organizations, especially in industries like construction, manufacturing, and healthcare, require their contractors and suppliers to have ISO 45001 certification. Small businesses with certification can access new contracts, tenders, and business opportunities that might have been unavailable otherwise.

6. Enhanced Internal Communication and Leadership Commitment

ISO 45001 emphasizes the importance of leadership commitment and clear communication in maintaining a successful OH&S management system. For small businesses:

- **Stronger leadership and management commitment**: The standard requires management to take a leading role in fostering a culture of safety. This commitment from leadership shows employees that their health and safety are a top priority, leading to increased respect and trust in the management.
- **Improved communication**: ISO 45001 improves communication by ensuring that safety policies, procedures, and roles are clearly defined and understood by everyone in the organization. Regular safety meetings, reports, and audits ensure that the management and staff are always aligned on safety goals.
- **Increased collaboration and teamwork**: The standard promotes collaboration among employees, management, and external stakeholders. Teams work together to identify hazards, assess risks, and create solutions, enhancing teamwork and mutual respect.

7. Continuous Improvement and Organizational Growth

ISO 45001 is built around the concept of continuous improvement. For small businesses, this means:

- **Ongoing improvements in OH&S performance**: The certification encourages businesses to continually assess their safety management systems, identify areas for improvement, and implement corrective actions. This ongoing process ensures that the organization remains proactive in managing risks and improving safety standards.
- **Adaptation to changes**: As laws, regulations, and business environments evolve, ISO 45001 helps businesses stay adaptable. Continuous improvement ensures that small businesses remain competitive, compliant, and able to address emerging risks.
- **Long-term business sustainability**: A commitment to OH&S leads to long-term sustainability. Fewer workplace accidents, improved employee

satisfaction, and better compliance help businesses thrive and grow while reducing operational risks.

8. Positive Impact on the Bottom Line

ISO 45001 helps businesses stay profitable by reducing the cost of workplace injuries, improving productivity, and enhancing overall operational efficiency. By investing in employee health and safety, small businesses benefit from:

- **Reduced absenteeism**: Fewer workplace injuries and illnesses mean that employees are absent less frequently, contributing to higher productivity and better overall performance.
- **Increased efficiency**: With fewer disruptions and a safer work environment, businesses can streamline operations, reduce downtime, and achieve better efficiency.
- **Stronger financial performance**: The combination of cost savings, improved employee productivity, and better market positioning leads to a stronger financial position, ultimately supporting growth and profitability.

Conclusion: Why ISO 45001 Certification Is Beneficial for Small Businesses

In conclusion, ISO 45001 certification offers small businesses a comprehensive framework to protect their employees, reduce risks, and comply with health and safety regulations. The benefits—ranging from improved safety, cost savings, and legal compliance to enhanced employee engagement and business reputation—make ISO 45001 a smart investment for any small enterprise. The certification not only ensures a safe and healthy workplace but also contributes to the long-term growth and success of the business.

Author's Personal Note

Over the years, I have worked with many organizations—from government institutions to small family-owned businesses—helping them implement international standards such as ISO 45001. One important observation I have consistently made is that workplace safety is not only a technical requirement; it is a reflection of leadership values and organizational culture.

Small enterprises often believe that international standards are designed mainly for large corporations with extensive resources. In reality, ISO 45001 can be even more valuable for smaller organizations. Small teams are closely connected, communication is faster, and leadership can directly influence workplace behavior. This creates a powerful opportunity to build a strong safety culture from the beginning.

During my consulting work across different countries, I have seen how a well-implemented Occupational Health and Safety Management System transforms businesses. It improves employee confidence, strengthens management responsibility, and creates a structured environment where risks are managed proactively rather than reactively.

For small enterprises, the real value of ISO 45001 lies not only in obtaining certification but in adopting a mindset of continuous improvement. Certification is simply a milestone.

My advice to every small business owner and manager reading this chapter is simple: start with commitment. Even small steps toward structured safety management can lead to significant improvements in employee well-being, operational stability, and long-term business success.

ISO 45001 is not merely a compliance tool—it is a strategic investment in people, trust, and sustainable growth.

— Dr. Mohamed-Ali Ibrahim

CHAPTER 2

The Certification Journey for Small Businesses

Achieving **ISO 45001:2018** certification is a structured and detailed process, especially for small businesses. While the journey may seem challenging, breaking it down into smaller, manageable steps can help small enterprises navigate the certification path effectively. This chapter outlines the key steps that small businesses must follow to achieve ISO 45001 certification, including the preparation, implementation, and continuous improvement stages.

Key Steps to Achieving Certification for Small Enterprises

1. Understanding ISO 45001:2018 Standards and Requirements

Before starting the certification journey, it is essential for small businesses to fully understand **ISO 45001:2018** and what it entails. This international standard provides a framework for managing Occupational Health and Safety (OH&S) and helps businesses identify hazards, assess risks, and improve safety performance.

Actions:

- **Study the ISO 45001 Standard**: Read and understand the core requirements of the ISO 45001:2018 standard. Identify the key principles, including based on based on risk assessment methods defined by the organization in line with ISO 45001:2018 Clause 6.1.2" to align with risk-based thinking., risk assessment, legal compliance, and leadership responsibilities.
- **Identify Scope and Relevance**: Determine how ISO 45001 applies to your specific business operations, industry, and regulatory requirements.

2. Conduct a Gap Analysis or Pre-Assessment

A **gap analysis** or **pre-assessment** is an essential first step for small businesses. It helps identify the gaps between the current OH&S practices in the organization and the requirements of ISO 45001. This process allows businesses to understand where improvements are needed to meet the standard.

Actions:

- **Review Existing OH&S Practices**: Evaluate the current OH&S system, including existing policies, procedures, and controls.
- **Identify Gaps**: Compare your existing OH&S system against the requirements of ISO 45001:2018. Identify the areas that need improvement or additional documentation.
- **Involve Key Stakeholders**: Engage management and relevant personnel in the gap analysis process to ensure all critical areas are reviewed.

3. Develop the OH&S Management System

After identifying gaps, the next step is to **develop** or **revise** the Occupational Health and Safety Management System (OHSMS) based on ISO 45001 requirements. This step involves creating policies, defining roles, and implementing systems that can effectively manage OH&S risks.

Actions:

- **Create an OH&S Policy**: Develop a comprehensive OH&S policy that aligns with ISO 45001 standards. This policy should reflect the organization's commitment to ensuring employee health and safety and must be endorsed by senior management.
- **Define Roles and Responsibilities**: Assign roles for managing safety and health within the organization. This includes appointing an OH&S manager or coordinator, if necessary, and establishing safety committees or teams.
- **Develop Procedures and Controls**: Create procedures to address key components such as based on based on risk assessment methods defined by the organization in line with ISO 45001:2018 Clause 6.1.2" to align with risk-based thinking., risk assessments, employee training, and incident reporting.

- **Establish Documentation**: ISO 45001 requires businesses to keep detailed records of their OH&S activities. This includes safety training logs, risk assessment documentation, and incident reports.

4. Implement the OH&S Management System

Once the OH&S management system is developed, the next step is to **implement** it across the organization. This phase involves communicating the system to employees, integrating it into daily operations, and ensuring it is functioning effectively.

Actions:

- **Communicate the System to Employees**: Ensure that all employees are informed about the new OH&S system, their responsibilities, and the importance of complying with safety procedures. This can be done through training sessions, meetings, and internal communications.
- **Provide Training**: Organize training programs to ensure employees are aware of health and safety risks and understand their roles in mitigating those

risks. Training should include emergency response protocols, safe work practices, and based on based on risk assessment methods defined by the organization in line with ISO 45001:2018 Clause 6.1.2" to align with risk-based thinking. techniques.

- **Monitor and Review**: Regularly monitor the system to ensure compliance. Track incidents, safety concerns, and employee feedback to identify any potential issues or areas for improvement.

5. Conduct Internal Audits

Internal audits are a critical part of the certification process. They help verify that the OH&S management system is functioning as intended and compliant with ISO 45001 standards. Internal audits also help identify areas for improvement before the formal external audit takes place.

Actions:

- **Plan and Schedule Audits**: Develop an audit schedule, ideally every six months or annually, depending on the business size and risk exposure.
- **Audit Performance**: Conduct internal audits to review the effectiveness of the OH&S system. This includes assessing risk management processes, incident records, training programs, and management reviews.
- **Identify Non-Conformities**: During the audit, note any non-conformities or areas where the OH&S system is not fully compliant with ISO 45001 requirements.
- **Implement Corrective Actions**: After the internal audit, take corrective actions to address any issues identified. This could involve revising procedures, improving training programs, or adjusting safety controls.

6. Engage a Certification Body for External Audit

After successfully conducting internal audits and addressing non-conformities, small businesses should engage an accredited certification body for the external certification audit. The external audit is a comprehensive evaluation by a third-party auditor to verify whether the business meets the ISO 45001:2018 standards.

Actions:

- **Select an Accredited Certification Body**: Choose an accredited certification body with experience in ISO 45001 audits. Ensure the body is recognized by national or international accreditation bodies.
- **Prepare for the External Audit**: Ensure that all documentation, policies, procedures, and records are in order and readily available for review by the external auditor.
- **Participate in the Audit**: During the audit, the certification body will review the OH&S management system, including policies, risk assessments, employee training records, and incident reports. They will also interview key staff members and observe workplace conditions.
- **Receive Audit Observations and recommendations must be documented; implementation is per organization discretion.**: The external auditor will provide Observations and recommendations must be documented; implementation is per organization discretion., including any non-conformities or areas for improvement. These must be addressed before certification is granted.

7. Certification Decision and Issuance

Based on the results of the external audit, the certification body will make a decision on whether the small business is eligible for ISO 45001 certification. If the business meets all the necessary requirements, the certification body will issue the certificate.

Actions:

- **Address Any Non-Conformities**: If the audit uncovers any issues, take corrective actions immediately. The certification body may require proof of these actions before issuing the final certificate.
- **Obtain Certification**: If the business passes the audit and complies with ISO 45001, the certification body will issue the ISO 45001:2018 certificate, formally recognizing the small business's commitment to occupational health and safety.

8. Continuous Improvement and Surveillance Audits

Achieving ISO 45001 certification is not the end of the journey. Continuous improvement is a fundamental aspect of the standard. Small businesses are required to maintain their OH&S system and undergo regular surveillance audits to ensure continued compliance.

Actions:

- **Monitor and Improve**: Regularly monitor the OH&S system to ensure it remains effective. Make improvements based on audit Observations and recommendations must be documented; implementation is per organization discretion., employee feedback, and changes in legal requirements.
- **Conduct Surveillance Audits**: Surveillance audits are conducted annually or every two years to ensure that the OH&S system remains compliant with ISO 45001 standards.
- **Maintain Certification**: To retain the ISO 45001 certification, small businesses must demonstrate continuous improvement in their OH&S practices during surveillance audits.

Conclusion

The journey to ISO 45001 certification for small businesses involves a structured and detailed process, but with proper preparation, it is achievable. The key steps—understanding the standard, developing and implementing an OH&S management system, conducting internal audits, and passing the external certification audit—are essential for small businesses to enhance safety, improve compliance, and achieve long-term success.

Role of Internal and External Audits for Small Enterprises

Internal and external audits are critical components of maintaining and achieving **ISO 45001:2018** certification for small businesses. These audits ensure that an Occupational Health and Safety (OH&S) management system is both compliant with ISO standards and effective in improving workplace safety. Both types of audits have distinct roles but work together to ensure continuous improvement and the overall success of the OH&S management system.

Below is a detailed explanation of the role of both **internal audits** and **external audits**, along with a table summarizing their differences and key contributions to the certification process.

1. Internal Audits for Small Enterprises

Internal audits are conducted by the organization itself, either by an internal team or a designated person responsible for evaluating the company's OH&S management system. The primary goal of internal audits is to assess whether the OH&S management system is effectively implemented, operating as intended, and adhering to ISO 45001:2018 requirements. This self-assessment ensures that any non-conformities or areas for improvement are identified before external auditors assess the business.

Role of Internal Audits:

- **Monitoring System Performance**: Internal audits provide an opportunity to regularly monitor the performance of the OH&S management system. This includes reviewing whether the established procedures and controls are being followed and identifying any areas where the system is not working as effectively as it should be.
- **Identifying Non-Conformities**: By performing internal audits, businesses can identify gaps in compliance or non-conformities in their OH&S system. These could include incomplete risk assessments, missing documentation, or ineffective safety measures.
- **Promoting Continuous Improvement**: Internal audits help drive a culture of continuous improvement by regularly assessing performance, identifying weaknesses, and suggesting corrective actions. These audits provide management with the feedback needed to improve the system's effectiveness.
- **Ensuring Compliance**: Internal audits ensure that the business is compliant with its own policies and procedures as well as with the ISO 45001 standard. It checks that any previously identified issues have been addressed and verifies that corrective actions have been taken.
- **Employee Engagement**: Internal audits can involve employees and ensure their participation in safety procedures. They create a feedback loop that encourages open communication between management and staff about safety concerns.

Key Tasks in Internal Audits:

- **Developing an Audit Schedule**: Small businesses should set up an audit schedule based on the complexity and risks of their operations. The audits can be scheduled quarterly, semi-annually, or annually depending on the business size.
- **Conducting Audits**: The internal audit team will assess various departments or functions, such as risk management, training programs, accident reporting, and emergency response plans.
- **Documenting Observations and recommendations must be documented; implementation is per organization discretion.**: Auditors will document their Observations and recommendations must be documented; implementation is per organization discretion. and identify any non-conformities, weaknesses, or opportunities for improvement.
- **Corrective Actions**: After the audit, small businesses are required to take corrective actions for identified non-conformities. This could involve revising procedures, providing additional training, or implementing better controls.

2. External Audits for Small Enterprises

External audits are conducted by an independent third-party certification body. These audits are part of the certification process and are required to achieve and maintain ISO 45001:2018 certification. The role of external audits is to evaluate whether the business is compliant with ISO 45001 standards, assess the effectiveness of the OH&S system, and provide a final certification decision.

Role of External Audits:

- **Objective Assessment**: External auditors provide an unbiased and objective assessment of the OH&S management system's compliance with ISO 45001:2018. They evaluate the effectiveness of the system and verify whether the organization's processes align with the standard's requirements.
- **Providing Certification**: External audits are crucial for obtaining and maintaining ISO 45001 certification. If the organization passes the audit, the certification body will grant the ISO 45001:2018 certification, recognizing the organization's commitment to health and safety.

- **Verifying Implementation**: External auditors verify that the OH&S management system is effectively implemented across the organization. They ensure that the system is not just documented but is also actively followed and functioning as intended.
- **Ensuring Global Standards Compliance**: Since ISO 45001 is internationally recognized, an external audit ensures that small enterprises meet global health and safety standards. This certification opens the door for businesses to compete in international markets and adhere to global best practices.
- **Surveillance Audits**: After obtaining certification, small businesses must undergo periodic surveillance audits, usually every 12 to 18 months. These audits check that the organization continues to comply with the standard and identify areas where improvements can be made.

Key Tasks in External Audits:

- **Preparation for External Audit**: Small businesses must prepare by ensuring that all documentation, records, and OH&S procedures are up to date and available for review by the external auditor.
- **On-Site Evaluation**: The external auditor will visit the business, review safety procedures, interview employees, and evaluate safety practices in the workplace. They may also assess management's involvement and commitment to OH&S.
- **Audit Report and Observations and recommendations must be documented; implementation is per organization discretion.**: After the audit, the external auditor will provide a detailed report of their Observations and recommendations must be documented; implementation is per organization discretion., including any non-conformities. If the system is fully compliant with ISO 45001:2018, they will issue the certification.
- **Certification Decision**: Based on the Observations and recommendations must be documented; implementation is per organization discretion., the certification body will make a final decision about whether to grant the certification. If the business does not meet the required standards, corrective actions will be required before certification is awarded.

3. Internal vs. External Audits for Small Enterprises

Aspect	Internal Audits	External Audits
Conducted by	Internal team or designated personnel	Independent, third-party certification body
Purpose	To monitor compliance with internal policies and ISO 45001	To assess overall compliance with ISO 45001 and issue certification
Frequency	Regular, typically quarterly or annually	Annually or bi-annually, depending on certification schedule
Focus	Identifying gaps, non-conformities, and areas for improvement	Verifying full compliance and system effectiveness
Outcome	Corrective actions for identified issues	ISO 45001 certification or non-conformity report
Engagement	Involves employees at all levels	Focuses on leadership and management review
Cost	Low cost, conducted in-house	Higher cost, paid to certification body

Conclusion

Both internal and external audits play a crucial role in helping small enterprises maintain effective Occupational Health and Safety (OH&S) management systems. Internal audits focus on continuous improvement and self-assessment, helping businesses identify issues and make corrective actions. External audits, on the other hand, provide an objective evaluation of the organization's adherence to ISO 45001 standards and lead to formal certification.

Small businesses must take both audits seriously, ensuring that they are continuously improving their OH&S systems and adhering to industry best practices. Regular internal audits prepare the organization for external audits, which help to secure ISO 45001 certification and sustain long-term safety improvements.

Time and Resource Management for Certification

Achieving **ISO 45001:2018** certification is a significant undertaking for any small business, requiring dedicated time, resources, and careful planning. Effective **time and resource management** are crucial to ensure that the process is not only successful but also efficient. For small businesses with limited staff and budgets, managing both time and resources effectively can be the difference between a smooth certification journey and a prolonged or costly process.

This section delves into the importance of managing time and resources throughout the certification process, offering a structured approach to ensure that small businesses stay on track and meet ISO 45001 requirements without overwhelming their operations.

1. Time Management for ISO 45001 Certification

Achieving ISO 45001 certification requires a structured, phased approach with careful planning to meet deadlines while ensuring compliance. Since small businesses often have fewer personnel, time management is especially important to avoid disruptions to everyday operations.

Key Steps for Effective Time Management:

A. Develop a Clear Timeline

A well-defined timeline with clear milestones ensures that the certification process progresses smoothly. Start by breaking the process down into phases (e.g., gap analysis, system development, training, internal audits, external audits, etc.).

Actions:

- **Set realistic deadlines**: Break the certification process into smaller tasks and set achievable deadlines for each. Be mindful of the business's operational schedule, holidays, and key business events to avoid interruptions.
- **Allocate time for training and engagement**: Provide sufficient time for employee training and internal communication to ensure everyone is aligned with the new OH&S practices.
- **Allow time for adjustments**: Flexibility is important. Unforeseen challenges may arise, so ensure that there is extra time built into the schedule to resolve issues without rushing.

B. Allocate Sufficient Time for Each Phase

The certification process is a multi-step journey. Each phase requires adequate time to complete effectively. Overly ambitious timelines may result in gaps or incomplete implementation, which could delay certification.

Actions:

- **Pre-certification phase**: Ensure that adequate time is allotted for the gap analysis and the development of the OH&S system. This phase may take several weeks or months, depending on the complexity of the business's operations.
- **Implementation phase**: Implementing the OH&S system across the business will require time for employee training, process development, and internal communication. Depending on the size of the workforce, this could take several weeks.
- **Audit preparation**: Prepare for internal and external audits by ensuring that documentation is complete, staff members are ready, and processes are well understood. This preparation phase can last up to a few weeks before the external audit.

C. Set Milestones and Track Progress

Establish clear milestones to track the progress of the certification process. Regular monitoring and reviews of the timeline will help ensure that deadlines are being met and corrective actions can be taken if progress slows down.

Actions:

- **Establish key milestones**: Examples of milestones include completing the gap analysis, completing risk assessments, conducting the first internal audit, and preparing for the external certification audit.
- **Monitor progress regularly**: Set weekly or bi-weekly check-ins to review the status of tasks and discuss any roadblocks.

D. Align Certification with Business Cycles

The certification process should align with the natural business cycle to minimize disruptions. This involves considering peak business periods, employee availability, and production schedules.

Actions:

- **Avoid busy seasons**: If the business has seasonal peaks, try to schedule the certification activities in the off-peak months when there's more time available.
- **Factor in employee schedules**: Plan for adequate staffing during key phases of the process, such as training or audits, to ensure that work does not fall behind during certification preparation.

2. Resource Management for ISO 45001 Certification

Effective resource management is essential for small businesses working on ISO 45001 certification. Resources include both human resources (staff time, expertise) and physical resources (technology, tools, external services). Proper management ensures that the business avoids over-committing and can allocate resources efficiently without impacting normal operations.

A. Assign Roles and Responsibilities

Identifying key roles and responsibilities for the certification process ensures that tasks are evenly distributed, and no one person or team is overwhelmed.

Actions:

- **Designate a project leader**: Appoint a project manager or OH&S coordinator responsible for overseeing the certification process. This individual will ensure that the timeline is followed, resources are allocated, and communication is smooth.
- **Involve key personnel**: Assign specific roles to employees in charge of risk assessments, training programs, documentation, and audits. This helps ensure that tasks are clearly defined and assigned to capable individuals.

- **Involve senior management**: Senior management should be involved in setting goals, approving resources, and demonstrating commitment to the OH&S system. Their support is vital for successful implementation.

B. Allocate Sufficient Budget for Certification

ISO 45001 certification involves certain financial costs, such as audit fees, certification body charges, training costs, and the cost of external consulting services (if needed). Properly managing the budget ensures the process stays within financial limits without sacrificing quality.

Actions:

- **Create a certification budget**: Estimate the costs of the entire certification process, including training programs, auditing fees, and any software or tools that may be needed for documenting the OH&S management system.
- **Monitor costs regularly**: Track the expenses during the certification process and adjust the budget if necessary. Consider potential cost-saving measures without compromising the quality of implementation.

C. Leverage External Support

Small businesses often benefit from external support, such as consultants, trainers, or safety experts, to help expedite the certification process. While hiring consultants or trainers incurs additional costs, they can provide valuable expertise and help streamline the implementation.

Actions:

- **Consult with experts**: Hiring an experienced ISO 45001 consultant can help ensure that the business's OH&S system is aligned with best practices. Consultants can also assist in preparing for audits and creating necessary documentation.
- **Invest in training**: Allocate resources to training employees on health and safety procedures and ISO 45001 standards. This investment enhances workforce competence and promotes employee buy-in to the new system.

D. Use Technology and Tools

Technology can streamline many processes involved in achieving ISO 45001 certification. From document management to incident reporting, software tools can help small businesses organize and manage the certification process more efficiently.

Actions:

- **Use OH&S management software**: Invest in software tools that can automate tasks such as risk assessments, incident reporting, and training records.
- **Leverage document management systems**: Use a cloud-based document management system to keep track of policies, procedures, and audit records. This simplifies the process of preparing for audits and ensuring compliance.

3. Balancing Resources and Time for Efficient Certification

For small enterprises, balancing the available resources and time is crucial to ensure that the certification process does not disrupt daily operations. Prioritizing resources based on business needs and focusing on key areas at different stages can lead to a more effective and streamlined certification journey.

Actions:

- **Start early**: Begin the certification process well in advance of any planned external audits. Give enough time for each phase of the process, including gap analysis, system development, employee training, and audit preparation.
- **Be realistic about capacity**: Understand that certification is a long-term commitment and allocate resources accordingly, making sure the business can still operate smoothly during the process.
- **Optimize resource allocation**: Focus resources on areas that will provide the most impact (such as risk assessments, employee training, and documentation management), while ensuring that less critical tasks don’t hinder progress.

Conclusion

Time and resource management are integral to successfully achieving ISO 45001 certification for small businesses. By planning carefully, allocating resources effectively, and ensuring that each phase of the certification process is given sufficient time and attention, small businesses can navigate the certification journey smoothly and efficiently. Good time and resource management not only ensures that the certification process is completed on schedule but also minimizes disruptions to daily operations, reduces costs, and increases the chances of a successful certification outcome.

Author's Personal Note

In my experience working with organizations around the world, the certification journey is often perceived as the most challenging part of implementing ISO standards. Many small business owners initially feel overwhelmed by the structured processes, documentation requirements, and audits involved in ISO 45001 certification.

However, what I have repeatedly observed is that once the journey begins, organizations quickly realize that the process itself brings clarity and structure to their operations. The steps described in this chapter—gap analysis, system development, employee training, internal audits, and certification—are not merely administrative tasks. They are tools that help organizations better understand their own processes, risks, and responsibilities.

Small enterprises, in particular, often progress faster than larger organizations because their decision-making processes are more agile and communication between employees and leadership is more direct. When leadership shows genuine commitment, employees naturally follow, and the entire organization begins to adopt a stronger culture of safety and responsibility.

Throughout my consulting and auditing career, I have seen many companies transform during the certification journey. What starts as a compliance project often evolves into a broader organizational improvement initiative. Teams become more organized, responsibilities become clearer, and employees gain confidence in their working environment.

The most important lesson I would like readers to take from this chapter is that certification should never be seen as the final objective. The real goal is building a sustainable management system that supports continuous improvement, protects employees, and strengthens the long-term stability of the organization.

Every successful certification journey begins with a simple but powerful decision: the commitment to improve.

— **Dr. Mohamed-Ali Ibrahim**

CHAPTER 3

The Certification Audit Process Explained for Small Enterprises

The **ISO 45001:2018** certification audit is the formal process by which a third-party certification body evaluates whether a business complies with the ISO 45001 standard for Occupational Health and Safety (OH&S). For small enterprises, understanding the structure and flow of this audit is crucial for a smooth and successful certification. The certification audit process ensures that businesses meet the requirements of the standard, and it is a significant milestone in achieving ISO 45001 certification.

This chapter provides a detailed explanation of the **structure of the ISO 45001 audit** for small businesses, including the audit stages, key activities, and roles of both the business and the auditor. By understanding this structure, small enterprises can prepare effectively for the audit, ensure compliance, and address any issues before the external audit is conducted.

The Structure of the ISO 45001 Audit for Small Businesses

The **ISO 45001 certification audit** process is typically divided into **two stages**: **Stage 1** (Documentation Review) and **Stage 2** (Certification Audit). Below is a breakdown of each stage and the steps involved in the certification audit process.

1. Stage 1: Documentation Review

The **Stage 1 audit** is primarily focused on reviewing the documentation of the OH&S management system (OHSMS) and determining whether the organization is ready for the full certification audit (Stage 2). This stage allows the auditor to assess whether the small business's OH&S system meets the minimum requirements of ISO 45001.

Key Activities in Stage 1:

A. Document Review

- The auditor will review key documentation to assess the compliance of the OH&S system with ISO 45001:2018 requirements. This includes:
  - **OH&S Policy**: A document outlining the business's commitment to health and safety.
  - **Risk Assessments**: Evaluating how hazards and risks are identified, assessed, and managed.
  - **Training and Competency Records**: Reviewing employee training and qualifications related to OH&S.
  - **Health and Safety Procedures**: Reviewing procedures for risk mitigation, emergency response, and reporting accidents.

B. Management System Review

- The auditor will check whether the OH&S management system is effectively integrated into the business and whether leadership and management are committed to maintaining and improving the system.
- This includes verifying whether the business has set OH&S objectives, defined roles and responsibilities, and allocated sufficient resources to implement the system.

C. Initial Gap Assessment

- The auditor will compare the business's current practices with ISO 45001 standards and identify any **gaps** that need to be addressed before moving to Stage 2. This is not an audit of effectiveness but rather an initial check to determine readiness.

D. Identification of Non-Conformities

- If the auditor identifies any major issues that may prevent the business from meeting ISO 45001 requirements, they will be flagged as **non-conformities**. These need to be resolved before Stage 2.

E. Preliminary Stage 2 Planning

- Based on the Stage 1 Observations and recommendations must be documented; implementation is per organization discretion., the auditor will propose a plan for the Stage 2 audit, which will focus on evaluating the implementation and effectiveness of the OH&S management system. The auditor may suggest improvements or modifications to the system before moving to Stage 2.

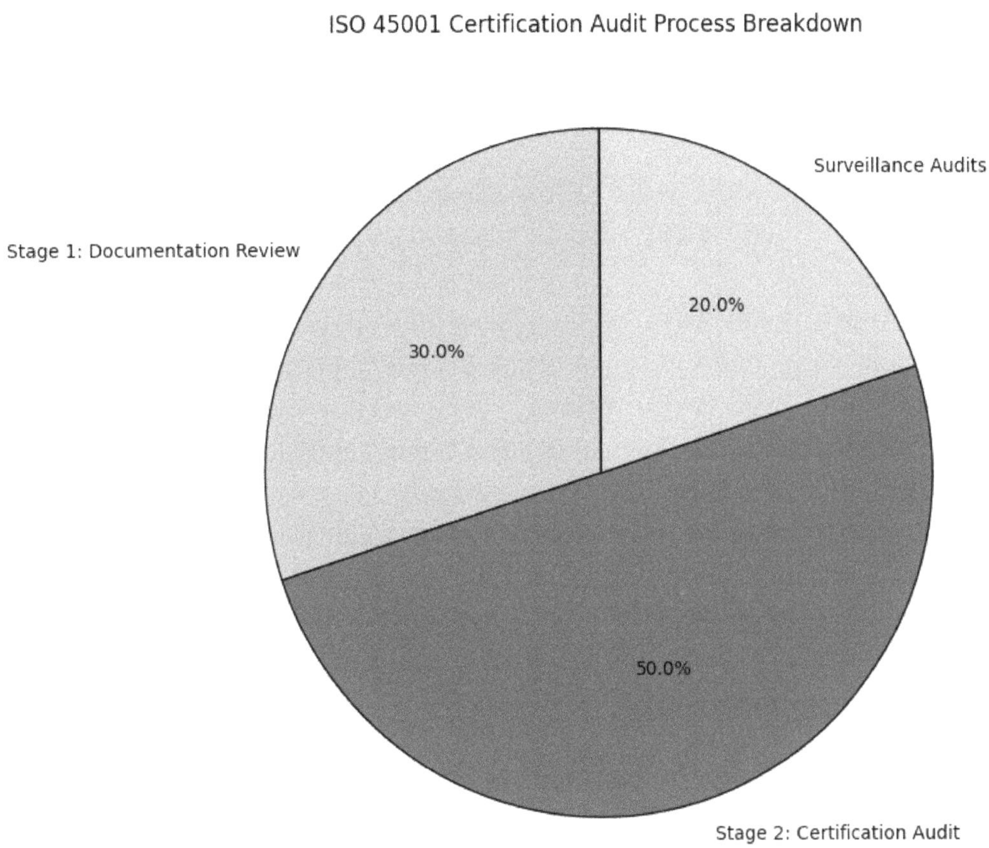

Outcome of Stage 1:

- **Readiness Assessment**: The business will receive feedback on whether it is ready for the full certification audit (Stage 2).
- **Non-Conformities**: If major non-conformities are found, the business will be required to address them before proceeding to Stage 2.
- **Recommendations**: The auditor may provide recommendations on improving the OH&S system and filling any gaps identified.

2. Stage 2: Certification Audit

Stage 2 is the **full certification audit**, where the auditor evaluates the implementation and effectiveness of the OH&S management system. This stage involves a thorough examination of how the business operates in practice and whether the system complies with all requirements of ISO 45001:2018.

Key Activities in Stage 2:

A. On-Site Audit

- The certification body will visit the small business's premises and conduct an on-site audit to assess the effectiveness of the implemented OH&S system. The auditor will:
    - **Interview employees**: To assess their awareness of the OH&S system, their roles, and the importance of safety procedures.
    - **Observe safety practices**: The auditor will observe actual work processes and safety practices to ensure they align with the documented procedures and ISO 45001 standards.
    - **Examine work environments**: The auditor will inspect the workplace to evaluate the physical safety measures in place (e.g., fire exits, equipment safety, PPE usage, emergency response plans).

B. Evaluating Compliance and Effectiveness

- The auditor will evaluate whether the OH&S system is effectively preventing incidents, minimizing risks, and ensuring employee health and safety. They will assess:
    - **Risk management processes**: Whether the business is actively identifying, assessing, and managing risks.
    - **Corrective and Preventive Actions**: Whether the business has implemented effective corrective and Preventive Actions to address non-conformities or safety incidents.
    - **Emergency preparedness**: How well the business is prepared to handle emergencies (e.g., fire drills, first-aid response, evacuation plans).
    - **Training and competency**: Whether employees are adequately trained and capable of performing their duties in a safe manner.

C. Reviewing Documentation and Records

- The auditor will ensure that the business maintains adequate records and documentation of its OH&S management system. This includes:
    - **Risk assessments**

- **Accident reports**
- **Training records**
- **Audit reports**
- **Corrective action logs**

D. Verification of Non-Conformities (if any from Stage 1)

- Any non-conformities identified during Stage 1 will be revisited and verified to ensure corrective actions have been taken before proceeding with certification.

E. Final Certification Decision

- Based on the Stage 2 audit Observations and recommendations must be documented; implementation is per organization discretion., the certification body will make a decision about whether the business meets the ISO 45001:2018 requirements. If the audit is successful, the certification body will issue the **ISO 45001:2018 certificate**.
- If non-conformities are found during Stage 2, the business will need to resolve them within a specified time frame and provide evidence of corrective actions before certification is granted.

Outcome of Stage 2:

- **Certification Granted**: If the business meets all requirements, ISO 45001:2018 certification will be awarded.
- **Non-Conformities Identified**: If there are any remaining non-conformities, the business will be given a specific period to address them and undergo a follow-up audit if necessary.

3. Post-Certification Surveillance Audits

After the certification is granted, small businesses must undergo **surveillance audits** (typically every 12 to 18 months) to ensure that the OH&S system is still effective and that the business continues to comply with ISO 45001:2018. These audits help maintain certification by confirming that the business is continually improving its OH&S management practices.

Summary Table: ISO 45001 Certification Audit Process

Audit Stage	Key Activities	Purpose
Stage 1: Documentation Review	- Review of OH&S policies, procedures, and risk assessments- Management system readiness check- Initial gap assessment- Preparation for Stage 2	To assess whether the business is ready for the full audit and to identify gaps in compliance
Stage 2: Certification Audit	- On-site audit of practices and processes- Interviews with employees- Inspection of work environments- Evaluation of risk management and corrective actions	To verify the effectiveness of the OH&S system and assess compliance with ISO 45001 standards
Surveillance Audits (Post-Certification)	- Ongoing monitoring of OH&S system implementation- Annual or bi-annual audits for continued compliance	To ensure the OH&S system remains effective and compliant with ISO 45001 standards

Conclusion

The **ISO 45001 certification audit** process for small businesses involves two key stages: Stage 1 (documentation review) and Stage 2 (the full certification audit). Each stage focuses on ensuring that the organization's OH&S system is in line with ISO 45001 standards. Internal and external audits work together to verify compliance, evaluate effectiveness, and ensure continuous improvement in workplace safety and health practices.

By understanding the structure and stages of the audit process, small businesses can better prepare for certification, address non-conformities, and maintain a safe and compliant work environment.

Preparing for the Certification Audit: A Guide for Small Enterprises

The **ISO 45001:2018 certification audit** is a key milestone for any small business that aims to formalize and improve its Occupational Health and Safety (OH&S) management system. This guide provides small enterprises with a comprehensive approach to prepare effectively for the audit. By following this guide, small businesses can ensure a smooth and successful audit process, leading to ISO 45001 certification and a safer, healthier workplace.

Why Preparing for the Certification Audit Is Crucial

The certification audit is the formal process through which a third-party auditor assesses whether a business's OH&S management system complies with the **ISO 45001:2018 standard**. For small businesses, proper preparation is essential for a few key reasons:

- **Avoiding Surprises**: Preparation helps ensure that there are no surprises during the audit, reducing stress and ensuring the audit runs smoothly.
- **Building Confidence**: When the team is well-prepared, employees and management alike will feel confident and ready for the external auditor's evaluation.
- **Maximizing Success**: Thorough preparation increases the chances of successfully obtaining ISO 45001 certification and meeting the auditor's expectations.
- **Identifying Gaps Early**: Proper preparation allows businesses to identify potential weaknesses in their OH&S systems before the auditor does, giving time to address them.

Key Steps to Prepare for the ISO 45001 Certification Audit

The preparation for the **ISO 45001 audit** involves multiple steps, ranging from reviewing and improving the existing OH&S system to ensuring that documentation is complete and that employees are trained. Below are the detailed steps that small businesses can take to prepare effectively for the certification audit.

1. Conduct a Pre-Audit or Gap Analysis

Before the external audit, small businesses should conduct an internal **gap analysis** or **pre-audit** to assess whether their current OH&S system meets the ISO 45001:2018 requirements.

Key Actions:

- **Assess current policies and procedures**: Review existing safety policies, procedures, and risk assessments to ensure they are aligned with ISO 45001.
- **Identify areas for improvement**: The gap analysis helps identify non-conformities or areas where the business is not yet meeting the ISO standard.
- **Engage employees**: Employees should be involved in the gap analysis process to identify any gaps from their perspective, as they are the ones most familiar with daily practices.

Benefits of Gap Analysis:

- It helps identify weaknesses in the current system before the audit.
- Allows time to implement corrective actions before the external auditor arrives.
- Ensures readiness for the external audit by addressing any non-conformities in advance.

2. Ensure Complete Documentation

One of the most important aspects of ISO 45001 certification is having comprehensive and up-to-date documentation. During the audit, the certification body will review various documents related to the OH&S system.

Key Documents to Prepare:

- **OH&S Policy**: A clear and concise policy that demonstrates commitment to employee health and safety.
- **Risk Assessment and Based on based on risk assessment methods defined by the organization in line with ISO 45001:2018 Clause 6.1.2" to align with risk-based thinking. Records**: Documentation showing that risks have been systematically assessed and managed.
- **Training and Competency Records**: Records proving that employees have been properly trained to work safely and understand their roles in the OH&S system.
- **Incident and Accident Reports**: Documentation showing how incidents and accidents are reported, investigated, and managed.

- **Corrective and Preventive Action Records**: Evidence of actions taken to address non-conformities or unsafe practices.

Tips for Document Preparation:

- **Ensure consistency**: Make sure that all OH&S documents are up-to-date and consistent with the actual practices in the workplace.
- **Organize documentation**: Keep documentation well-organized and easily accessible. This will make it easier for the auditor to review them during the audit.
- **Review for completeness**: Ensure all required documents are complete and readily available to demonstrate compliance.

3. Engage and Train Employees

An effective OH&S management system requires the active participation of all employees, not just management. Therefore, training and engaging employees in the preparation process is essential.

Key Actions:

- **Safety training**: Provide training to employees on the OH&S system, their roles and responsibilities, risk management, emergency procedures, and any new safety protocols.
- **Employee awareness**: Make sure that all employees understand the importance of ISO 45001 certification and how it impacts their health and safety at work.
- **Engage employees in safety culture**: Encourage employees to participate in safety committees or provide feedback on existing safety practices. Their involvement is essential to maintaining a safety-conscious culture.

Benefits of Employee Engagement:

- It helps ensure that all employees are familiar with OH&S policies and practices.
- It boosts morale, as employees feel more responsible and engaged in creating a safer work environment.

- Employee participation can highlight potential issues or risks that management may have missed.

4. Conduct Internal Audits

An **internal audit** is a critical part of preparing for the external certification audit. Internal audits are a self-assessment tool that helps businesses evaluate whether their OH&S management system is functioning effectively and meeting ISO 45001 standards.

Key Actions:

- **Audit the OH&S system**: Perform audits across all aspects of the OH&S system, such as risk assessments, employee training, and emergency procedures.
- **Identify non-conformities**: During the audit, look for areas where the business may not be in full compliance with ISO 45001. These issues should be addressed before the external audit.
- **Corrective actions**: If non-conformities are found during the internal audit, take corrective actions to address them before the external audit.

Benefits of Internal Audits:

- It ensures that the system is operating effectively and highlights areas of weakness.
- It provides an opportunity to resolve issues internally before the external audit.
- Regular internal audits foster a culture of continuous improvement and ensure ongoing compliance.

5. Simulate the External Audit (Mock Audit)

A **mock audit** is a dry run of the external certification audit. It involves conducting a simulated audit to identify any issues that may arise during the actual audit. This helps the business to become familiar with the process and identify any last-minute corrections that need to be made.

Key Actions:

- **Conduct a mock audit**: Assign internal auditors or external consultants to conduct the mock audit, following the same procedures as the external auditor would.
- **Evaluate readiness**: Use the mock audit to evaluate the business's preparedness for the external audit, focusing on documentation, practices, and employee engagement.
- **Correct any issues**: If any non-conformities or areas for improvement are identified, take corrective actions immediately.

Benefits of a Mock Audit:

- It helps identify potential problems before the official audit, giving the business time to resolve them.
- It boosts confidence by allowing the business to practice the process in a low-pressure setting.
- It provides insight into the external auditor's perspective and helps ensure that all requirements are met.

6. Final Preparation Before the External Audit

In the final stages of preparation, small businesses need to ensure that all the necessary elements are in place before the external auditor arrives. This includes reviewing and ensuring the effectiveness of the OH&S system, finalizing any last-minute changes, and ensuring that all documentation is ready for review.

Key Actions:

- **Review audit readiness**: Ensure that all documents are complete, the OH&S system is implemented, and employees are ready to engage with the auditors.
- **Ensure management availability**: Make sure that senior management is available during the external audit to demonstrate their commitment to the OH&S system.
- **Confirm employee participation**: Employees should be available to answer questions from the auditor and demonstrate their knowledge of safety protocols.

Conclusion

Preparation for the ISO 45001 certification audit is a detailed and structured process that requires commitment, planning, and collaboration. By conducting a gap analysis, ensuring complete documentation, training employees, performing internal audits, and conducting mock audits, small enterprises can increase their chances of a successful certification audit. Proper preparation not only ensures a smooth audit process but also improves the overall effectiveness of the OH&S management system, creating a safer workplace for employees and a more compliant organization.

Roles and Responsibilities During the Audit in Small Enterprises

The **ISO 45001:2018** certification audit process involves a collaborative effort between the business and the certification body. Both internal stakeholders (such as management and employees) and external auditors play essential roles in ensuring a smooth and successful audit. Understanding the specific roles and responsibilities of each party is crucial for small enterprises to prepare adequately for the audit.

This section provides a detailed explanation of the various roles and responsibilities involved during the audit, focusing on what is expected from management, employees, and external auditors. By understanding these roles, small businesses can ensure a structured and organized audit process, which increases the likelihood of a successful ISO 45001 certification.

1. Roles and Responsibilities of Management During the Audit

The role of **management** is vital in ensuring that the organization is prepared for the ISO 45001 audit and that the OH&S management system operates effectively. Senior management is responsible for setting the direction, providing resources, and demonstrating leadership throughout the process.

Key Responsibilities of Management:

A. Demonstrating Leadership and Commitment

- **Commitment to Safety**: Management must demonstrate a clear commitment to health and safety at all levels of the organization. This includes leading by example, prioritizing safety, and allocating necessary resources for OH&S initiatives.
- **Oversight of OH&S System**: Senior management should oversee the implementation and continuous improvement of the OH&S management system, ensuring it is integrated into the organization's broader management practices.
- **Engagement with Auditors**: During the audit, management must engage with the external auditors, answering questions about the OH&S system, leadership commitment, and business practices. This involvement helps to showcase management's active role in safety initiatives.

B. Providing Resources and Support

- **Resource Allocation**: Management is responsible for ensuring the organization has sufficient resources, including personnel, time, and financial support, to implement and maintain the OH&S system. This may include investing in safety equipment, training programs, or consultancy services.
- **Staff Engagement**: Management should support employees by ensuring they are trained, informed, and fully engaged in the health and safety initiatives.

C. Establishing Clear Safety Objectives and Performance Indicators

- **Setting Clear Objectives**: Management should set clear, measurable OH&S objectives in alignment with ISO 45001. These objectives must be communicated to all employees and linked to the organization's overall business goals.
- **Monitoring and Reviewing Performance**: Management must regularly review the effectiveness of the OH&S system and make adjustments as needed to improve safety performance.

D. Addressing Non-Conformities and Corrective Actions

- **Non-Conformity Management**: If the external auditor identifies any non-conformities during the audit, management is responsible for overseeing corrective actions. This includes making necessary changes to improve processes, policies, or procedures to meet ISO 45001 standards.
- **Corrective Action Plans**: After the audit, management should lead the effort to develop corrective action plans and ensure that they are implemented effectively.

2. Roles and Responsibilities of Employees During the Audit

Employees are central to the success of the **ISO 45001:2018** certification process. While management plays a leadership role, employees are responsible for implementing safety procedures, following best practices, and participating in the audit process.

Key Responsibilities of Employees:

A. Understanding OH&S Policies and Procedures

- **Knowledge of Safety Practices**: Employees must be familiar with the organization's OH&S policies, procedures, and safety practices. This includes understanding their individual roles and responsibilities in maintaining a safe work environment.
- **Compliance with Safety Protocols**: Employees are expected to follow all safety protocols, including risk assessment procedures, emergency response protocols, and accident reporting procedures.

B. Participating in the Audit Process

- **Engagement During the Audit**: Employees will likely be interviewed by the external auditors to verify that they are knowledgeable about the company's safety practices. They must be ready to answer questions about how they apply the safety procedures in their day-to-day work.
- **Providing Evidence of Safety Practices**: Employees should be prepared to demonstrate the implementation of safety practices. This could include

showing records of safety training, providing examples of risk assessments, or explaining how they report hazards.

C. Reporting Issues and Non-Conformities

- **Open Communication**: Employees must be willing to report any safety concerns, incidents, or non-conformities they have observed in the workplace. Transparent reporting helps improve the OH&S system and ensures the business remains compliant with ISO 45001.
- **Feedback for Improvement**: Employees should provide feedback on how safety practices could be improved or made more effective. Employee involvement in the process ensures the OH&S system is continuously enhanced.

3. Roles and Responsibilities of the External Auditors

External auditors are independent professionals or teams hired from an accredited certification body to evaluate the organization's compliance with the ISO 45001:2018 standard. The role of external auditors is critical in verifying the business's adherence to the standard and determining whether it qualifies for ISO certification.

Key Responsibilities of External Auditors:

A. Conducting the Audit

- **Document Review**: External auditors will review key OH&S documents, such as risk assessments, safety policies, employee training records, and incident reports. They will check whether these documents meet ISO 45001 requirements.
- **On-Site Assessment**: Auditors will visit the workplace to observe safety practices, interview employees, and assess the work environment. They will verify that the practices match the documented policies and procedures.
- **Objective Assessment**: Auditors conduct their evaluation impartially and objectively. They ensure that the OH&S management system is operating effectively, identifying any gaps or areas of non-compliance.

B. Providing Audit Observations and recommendations must be documented; implementation is per organization discretion. and Reports

- **Non-Conformities**: If the auditor finds areas where the OH&S management system does not comply with ISO 45001, they will document these as **non-conformities**. The auditor will also suggest corrective actions that the business must take.
- **Audit Report**: After the audit, the external auditor will prepare a detailed audit report, summarizing their Observations and recommendations must be documented; implementation is per organization discretion., including non-conformities, strengths of the system, and areas for improvement. The final report will also include a recommendation on whether the business qualifies for ISO 45001 certification.
- **Certification Decision**: Based on the audit Observations and recommendations must be documented; implementation is per organization discretion., the external auditor will make the final decision on whether the business meets the ISO 45001 requirements and should be awarded the certification.

C. Offering Guidance for Improvement

- **Best Practice Recommendations**: While the external auditor's role is to assess compliance, they may also offer recommendations on improving the OH&S management system, based on their experience with other organizations.
- **Support for Corrective Actions**: In cases of non-conformity, auditors may provide advice on how to implement corrective actions effectively.

4. Summary Table: Roles and Responsibilities During the Audit

Role	Responsibilities
Management	- Demonstrate leadership and commitment to OH&S- Provide resources and support- Set safety objectives- Address non-conformities and corrective actions

Employees	- Understand and comply with safety practices- Engage in the audit process (interviews, providing evidence)- Report issues and participate in continuous improvement
External Auditors	- Conduct the audit and assess compliance- Review documentation and on-site practices- Identify non-conformities and provide recommendations for improvement

Conclusion

The **ISO 45001 certification audit** involves the active participation of management, employees, and external auditors. Management plays a leadership role in driving safety culture and ensuring resources are available, while employees implement safety practices and contribute to the audit process. External auditors evaluate compliance and offer recommendations for continuous improvement. By understanding and fulfilling their roles and responsibilities, small businesses can ensure a smooth audit process, improve workplace safety, and achieve ISO 45001 certification successfully.

Author's Personal Note

Throughout my years of conducting and leading certification audits in different industries and countries, I have noticed that the word "audit" often creates unnecessary anxiety within organizations. Many teams prepare for audits as if they are facing an inspection designed to expose mistakes. In reality, the purpose of an ISO audit is very different.

An audit is primarily a structured dialogue between the organization and the auditor. It is an opportunity to verify that the systems designed to protect employees and manage risks are functioning effectively. When organizations understand this perspective, the audit process becomes much less intimidating and significantly more constructive.

In my professional experience as a lead auditor, I have seen that the most successful organizations are not necessarily those with the most complex documentation, but those with a genuine culture of safety. When employees clearly understand procedures, when management demonstrates visible commitment, and when safety practices are consistently applied in daily operations, the audit process becomes a natural confirmation of what the organization already practices.

Small enterprises often perform exceptionally well during audits because communication within their teams is more direct and responsibilities are clearly understood. When employees are engaged and management is committed, auditors quickly recognize the authenticity of the system.

It is important to remember that an audit is not designed to punish organizations but to help them strengthen their management systems. Non-conformities, when identified, should be seen as opportunities for improvement rather than failures.

Organizations that approach audits with transparency, preparation, and openness to learning often gain the greatest value from the process. In the long run, these audits help build stronger, safer, and more resilient workplaces.

— **Dr. Mohamed-Ali Ibrahim**

CHAPTER 4

Key Requirements for Certification in Small Enterprises

Core ISO 45001 Requirements and Compliance for Small Businesses

Achieving **ISO 45001:2018** certification requires small businesses to meet a set of core requirements designed to ensure the effective management of Occupational Health and Safety (OH&S). These requirements not only help protect employees but also improve overall business efficiency, reduce risks, and ensure compliance with legal and regulatory standards.

In this chapter, we will explore the core requirements of ISO 45001 and how small businesses can implement them to achieve compliance. These requirements serve as the foundation for creating a strong OH&S management system that leads to safer workplaces, enhanced employee morale, and greater operational efficiency.

1. Context of the Organization

What It Means:

ISO 45001 requires small businesses to first understand the context in which they operate and how OH&S management integrates with their overall organizational goals. This involves analyzing both internal and external factors that may affect employee health and safety.

Key Actions for Compliance:

- **Identify Internal and External Factors**: Determine factors that could influence your OH&S system. These may include the organizational culture, structure, operational processes, legal obligations, and market conditions.

- **Understand Stakeholder Needs**: Identify and understand the needs and expectations of interested parties (e.g., employees, contractors, suppliers, regulatory authorities). Ensure these needs are considered when designing your OH&S management system.
- **Establish Boundaries**: Define the scope of the OH&S management system. Ensure it covers all aspects of the organization that impact employee safety.

Why It's Important:

Understanding the organization's context ensures that the OH&S system is relevant to the business's needs and challenges. It also helps to align safety initiatives with broader organizational goals.

2. Leadership and Worker Participation

What It Means:

Leadership plays a crucial role in ISO 45001 compliance. The standard requires that top management demonstrate their commitment to OH&S and actively engage with employees at all levels to create a safety-conscious culture.

Key Actions for Compliance:

- **Leadership Commitment**: Senior management must demonstrate clear leadership by providing the necessary resources, support, and authority to implement and maintain the OH&S system.
- **Establish OH&S Policy and Objectives**: Management must develop and implement an OH&S policy that aligns with the business's objectives. The policy should define goals for improving safety, compliance, and health performance.
- **Employee Engagement**: Actively involve employees in safety-related activities, decision-making, and continuous improvement. Employees should be encouraged to participate in safety committees, based on based on risk assessment methods defined by the organization in line with ISO 45001:2018 Clause 6.1.2" to align with risk-based thinking., and risk assessment.

- **Clear Communication**: Ensure that OH&S information is communicated throughout the organization, and employees are aware of their roles, responsibilities, and the importance of health and safety measures.

Why It's Important:

Active leadership commitment and employee participation are essential for building a strong safety culture. When management demonstrates commitment to safety, it encourages employees to take responsibility for their own well-being and that of their colleagues.

3. Planning for OH&S Risks and Opportunities

What It Means:

The standard emphasizes a **proactive approach** to identifying and managing risks. Small businesses must systematically assess risks to Workers' health and safety and take steps to eliminate or mitigate these risks.

Key Actions for Compliance:

- **Based on based on risk assessment methods defined by the organization in line with ISO 45001:2018 Clause 6.1.2" to align with risk-based thinking. and Risk Assessment**: Regularly identify potential hazards and assess associated risks. Use methods such as risk matrices or based on based on risk assessment methods defined by the organization in line with ISO 45001:2018 Clause 6.1.2" to align with risk-based thinking. tools to assess severity and likelihood.
- **Legal and Other Requirements**: Ensure compliance with legal requirements related to OH&S, including national and international regulations. Regularly monitor changes in relevant legislation and integrate them into your safety management practices.
- **Establish OH&S Objectives**: Based on the identified risks, establish specific, measurable, achievable, relevant, and time-bound (SMART) OH&S objectives that align with organizational goals.
- **Address Opportunities for Improvement**: Look for opportunities to improve safety practices, reduce risk, and promote worker health. For

example, this could include improving training programs, introducing new safety technologies, or refining emergency response procedures.

Why It's Important:

By identifying and addressing OH&S risks and opportunities, businesses can prevent incidents, reduce the cost of accidents, and continuously improve their safety practices. Proactive risk management enhances workplace safety and employee well-being.

4. Support: Resources, Competence, and Awareness

What It Means:

ISO 45001 requires businesses to provide the necessary resources and competencies to implement and maintain their OH&S management system effectively. This includes training, awareness programs, and the provision of tools and equipment.

Key Actions for Compliance:

- **Provide Resources**: Ensure that the business has sufficient resources, including financial, human, and technological support, to implement the OH&S system effectively.
- **Ensure Competence**: Provide training to employees to ensure they have the necessary skills and knowledge to perform their jobs safely. This includes safety training, risk assessment workshops, and emergency response drills.
- **Create Awareness**: Increase awareness among employees about OH&S risks, their roles in maintaining safety, and the importance of compliance with safety policies and procedures.

Why It's Important:

Providing adequate resources and training ensures that the OH&S management system functions effectively and employees can operate safely. It also ensures that small businesses meet their obligations under ISO 45001 and minimize the risk of incidents.

5. Operation: Managing Risks and Controls

What It Means:

The standard requires businesses to establish processes to manage the identified risks and ensure that safety controls are in place. This includes procedures for dealing with emergency situations and monitoring the effectiveness of risk controls.

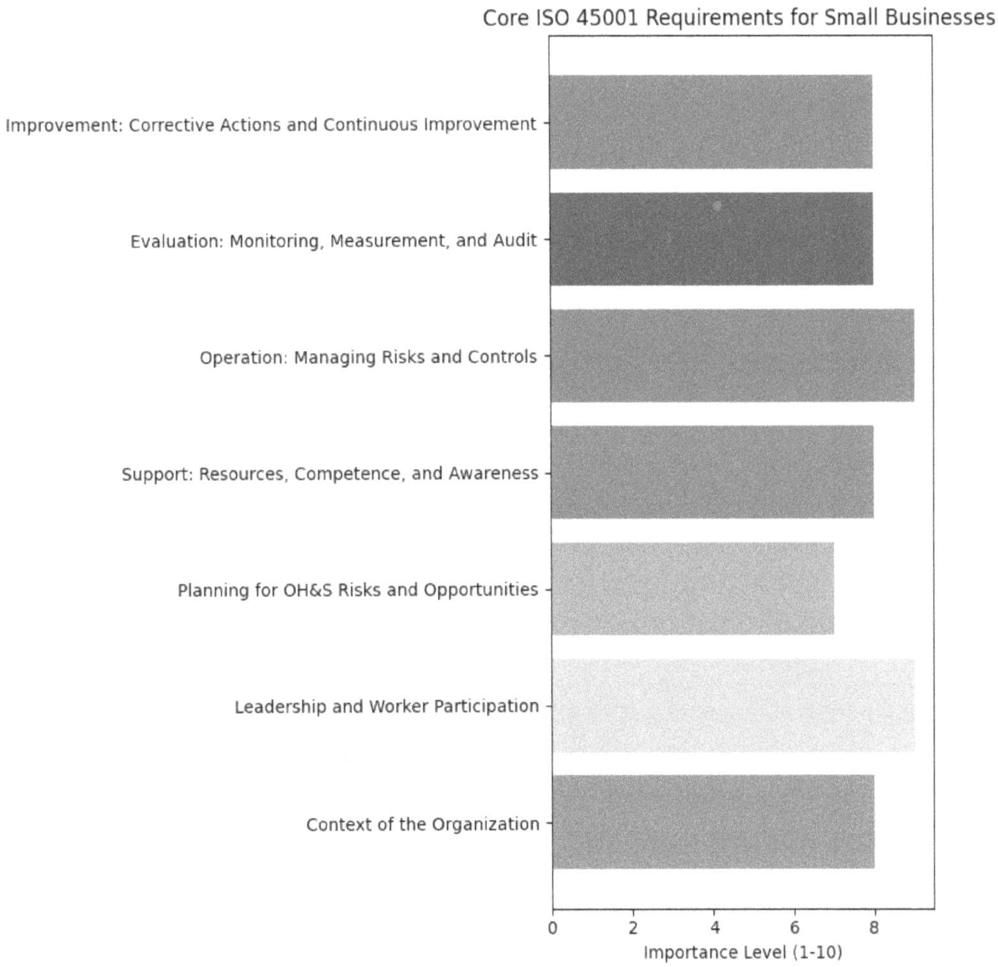

Key Actions for Compliance:

- **Implement Control Measures**: Put in place measures to control or eliminate identified risks. These may include engineering controls (e.g., ventilation systems), administrative controls (e.g., job rotation), or personal protective equipment (PPE).

- **Prepare for Emergencies**: Develop emergency response plans to address potential crises, such as accidents, fires, or health hazards. Regularly conduct drills to ensure readiness.
- **Monitor and Measure Safety Performance**: Regularly monitor safety performance, including tracking incidents, near misses, and safety audits. This will help identify areas for improvement and ensure compliance with safety objectives.

Why It's Important:

Managing risks and establishing controls are vital for preventing incidents and maintaining a safe workplace. By implementing robust controls, small businesses can ensure that health and safety hazards are mitigated.

6. Evaluation: Monitoring, Measurement, and Audit

What It Means:

The business must regularly evaluate the performance of its OH&S management system to ensure it is functioning effectively. This includes tracking progress against safety objectives and performing internal audits.

Key Actions for Compliance:

- **Monitor Performance**: Use metrics and indicators to track safety performance and measure the effectiveness of safety practices. This could include tracking accident rates, training completion, or compliance with safety procedures.
- **Conduct Internal Audits**: Regular internal audits help ensure that the OH&S management system is operating as intended. Audits should assess compliance with ISO 45001 and identify any areas that need improvement.
- **Management Review**: Senior management should review safety performance regularly to ensure continuous improvement. This involves assessing audit results, incident reports, and safety performance data.

Why It's Important:

Ongoing evaluation ensures that the OH&S system remains effective and relevant. It helps identify areas for improvement and ensures that the business is continually meeting the requirements of ISO 45001.

7. Improvement: Corrective Actions and Continuous Improvement

What It Means:

ISO 45001 emphasizes the importance of continuous improvement. Small businesses must address non-conformities and take corrective actions to improve the OH&S management system over time.

Key Actions for Compliance:

- **Corrective Actions**: When non-conformities are identified—whether through audits, inspections, or incident reports—small businesses must take corrective actions to address them. These actions should prevent the same issues from recurring.
- **Preventive Actions**: Businesses should also identify and address potential issues before they occur. This could involve refining processes, introducing new safety measures, or enhancing training programs.
- **Consider opportunities for improvement**: Continuously seek ways to improve the OH&S system, whether through new technologies, improved practices, or employee-driven safety initiatives.

Why It's Important:

Corrective and Preventive Actions ensure that the OH&S system is constantly evolving and improving. This leads to a more effective safety management system and a healthier, safer work environment.

Conclusion: Compliance with ISO 45001

The **core requirements** of ISO 45001:2018 provide a comprehensive framework for small businesses to improve their OH&S management system. By addressing areas such as leadership, planning, risk management, employee engagement, and continuous improvement, small enterprises can create a safe and healthy work environment that meets both regulatory requirements and international safety standards. Compliance with ISO 45001 not only protects employees but also improves overall business performance and enhances the company's reputation in the marketplace.

Essential Documentation and Evidence for Small Enterprises

Achieving **ISO 45001:2018** certification requires small businesses to present key **documentation** and **evidence** during the certification audit. This documentation serves as proof that the organization has implemented and is maintaining an effective **Occupational Health and Safety (OH&S) management system** that complies with ISO 45001 standards.

In this section, we will explain the **essential documentation** that small businesses must have in place for the audit and the **focus areas** that auditors will pay attention to during the certification audit.

1. Essential Documentation for ISO 45001 Certification

ISO 45001:2018 requires businesses to establish, maintain, and review specific documents to ensure compliance with the standard. The certification audit evaluates whether these documents are comprehensive, accurate, and reflective of the organization's practices. Below is a list of essential documents that small businesses should prepare for the audit.

A. OH&S Policy

- **What it is**: A documented statement that outlines the organization's commitment to protecting the health and safety of employees, contractors, and visitors. The policy must align with ISO 45001:2018 and set the tone for the organization's safety culture.
- **Audit Focus**: The auditor will check whether the OH&S policy is in place, whether it's communicated to employees, and whether it's consistent with the organization's goals.

B. Risk Assessment and Based on based on risk assessment methods defined by the organization in line with ISO 45001:2018 Clause 6.1.2" to align with risk-based thinking.

- **What it is**: A critical part of the OH&S system, this document outlines the process for identifying workplace hazards and assessing the risks associated with them. This includes physical, chemical, biological, ergonomic, and psychosocial risks.
- **Audit Focus**: Auditors will examine how risks are identified, the criteria for risk assessment, and whether appropriate control measures have been implemented. The auditor will also verify if the risk assessments are regularly reviewed and updated.

C. Training and Competency Records

- **What it is**: Documents that demonstrate the training, skills, and qualifications of employees regarding OH&S procedures. These records should show that employees are competent to handle workplace hazards and comply with safety protocols.
- **Audit Focus**: Auditors will check whether employees have received adequate training for their roles and responsibilities, whether training records are up to date, and whether employees are aware of their safety duties.

D. Incident and Accident Reports

- **What it is**: Records of accidents, near misses, injuries, and illnesses that have occurred in the workplace. These reports should include details on the causes, actions taken, and any corrective measures implemented.
- **Audit Focus**: The auditor will review accident and incident records to evaluate the effectiveness of the reporting and investigation procedures. The auditor will also check if corrective actions were taken in response to incidents.

E. Corrective and Preventive Action Records

- **What it is**: Documentation of actions taken to correct or prevent non-conformities and unsafe practices. This includes corrective actions for

identified issues, preventive measures to avoid future incidents, and follow-up activities.
- **Audit Focus**: Auditors will evaluate how non-conformities are managed and whether the corrective actions are effectively addressing the root causes. They will also check whether preventive measures are in place and being followed.

F. Internal Audit Reports

- **What it is**: Reports generated from internal audits that assess whether the OH&S management system is operating as intended. Internal audits identify gaps in compliance and areas for improvement.
- **Audit Focus**: External auditors will review internal audit records to verify that the organization is regularly evaluating its OH&S system, that non-conformities have been identified, and that corrective actions have been taken.

G. Emergency Response Plans

- **What it is**: Documented procedures for responding to emergencies such as fires, chemical spills, medical emergencies, or natural disasters. The plans should detail the roles and responsibilities of employees during an emergency, along with evacuation routes and emergency contacts.
- **Audit Focus**: The auditor will assess whether emergency plans are comprehensive, communicated to employees, and practiced through drills. The auditor will also ensure that emergency response procedures align with actual workplace hazards.

2. Focus Areas for Small Enterprises During the Audit

The **ISO 45001 certification audit** is a comprehensive process that evaluates the effectiveness of the OH&S management system. Auditors will focus on several key areas to determine whether the business meets the requirements of the standard. Below are the primary focus areas during the audit for small enterprises.

A. Leadership and Commitment

Focus Area: The auditor will evaluate the level of leadership commitment to OH&S management, including whether senior management is actively involved in safety initiatives, provides necessary resources, and communicates the importance of safety throughout the organization.

Key Questions the Auditor Will Ask:

- How does leadership demonstrate its commitment to health and safety?
- Are OH&S objectives aligned with the overall business goals?
- Is there adequate resource allocation for safety initiatives?

B. Employee Involvement and Participation

Focus Area: Employee participation is crucial for maintaining an effective OH&S system. Auditors will assess whether employees are engaged in safety-related activities, such as risk assessments, safety committees, incident reporting, and continuous improvement efforts.

Key Questions the Auditor Will Ask:

- How are employees involved in based on based on risk assessment methods defined by the organization in line with ISO 45001:2018 Clause 6.1.2" to align with risk-based thinking. and risk assessments?
- Are there mechanisms for employees to report safety concerns or suggestions for improvement?
- Are safety meetings and committees actively engaged in improving OH&S practices?

C. Based on based on risk assessment methods defined by the organization in line with ISO 45001:2018 Clause 6.1.2" to align with risk-based thinking. and Risk Assessment

Focus Area: One of the key requirements of ISO 45001 is identifying hazards and assessing the associated risks. Auditors will review the process used by the business to identify potential hazards, assess risks, and implement control measures.

Key Questions the Auditor Will Ask:

- How are workplace hazards identified and assessed?

- Are risk assessments documented, and are they regularly reviewed?
- Are appropriate control measures implemented to mitigate identified risks?

D. Legal Compliance

Focus Area: ISO 45001 requires that businesses comply with relevant national and international safety regulations. Auditors will check whether the small enterprise has identified applicable legal requirements and is actively ensuring compliance with them.

Key Questions the Auditor Will Ask:

- How does the business ensure compliance with local and international OH&S regulations?
- Is there a process for staying up to date with changes in safety laws and regulations?
- Are compliance records maintained and regularly reviewed?

E. Effectiveness of the OH&S Management System

Focus Area: Auditors will assess whether the OH&S management system is operating as intended, whether it is achieving the set OH&S objectives, and whether it is continuously improving.

Key Questions the Auditor Will Ask:

- How does the business monitor the effectiveness of its OH&S management system?
- Are safety objectives being met, and how is performance measured?
- What actions are taken when performance does not meet expectations?

F. Corrective and Preventive Actions

Focus Area: Corrective actions are crucial for addressing non-conformities and unsafe practices. Auditors will focus on how corrective and Preventive Actions are identified, implemented, and followed up on.

Key Questions the Auditor Will Ask:

- How are non-conformities tracked and addressed?
- Are corrective actions documented and implemented in a timely manner?
- Are Preventive Actions taken to eliminate potential risks before they occur?

G. Emergency Preparedness and Response

Focus Area: Emergency preparedness is a vital part of workplace safety. Auditors will evaluate whether the business has effective emergency response plans in place and whether employees are trained to respond to emergencies.

Key Questions the Auditor Will Ask:

- Does the business have an emergency response plan in place?
- Are employees trained on emergency procedures?
- Are emergency drills conducted regularly?

Summary Table: Essential Documentation and Focus Areas for the ISO 45001 Audit

Documentation	Focus Area	Audit Focus
OH&S Policy	Leadership and Commitment	Assess management's involvement and commitment to safety
Risk Assessment and Based on based on risk assessment methods defined by the organization in line with ISO 45001:2018 Clause 6.1.2" to align with risk-based thinking.	Employee Involvement and Participation	Evaluate how employees participate in based on based on risk assessment methods defined by the organization in line with ISO 45001:2018 Clause 6.1.2" to align with risk-based thinking. and reporting
Training and Competency Records	Based on based on risk assessment methods defined by the organization in line with ISO 45001:2018 Clause 6.1.2" to align with risk-based thinking. and Risk Assessment	Review the process for identifying and mitigating risks
Incident and Accident Reports	Legal Compliance	Ensure compliance with legal and regulatory requirements

Corrective and Preventive Action Records	Effectiveness of the OH&S Management System	Verify that corrective actions are taken and system performance is measured
Internal Audit Reports	Corrective and Preventive Actions	Check if corrective and Preventive Actions are implemented effectively
Emergency Response Plans	Emergency Preparedness and Response	Assess the effectiveness of emergency preparedness and employee readiness

Conclusion

The ISO 45001 certification audit for small businesses requires comprehensive documentation and effective systems to meet the key areas of focus outlined above. By ensuring that all necessary documentation is in place and addressing the focus areas during the audit, small businesses can demonstrate their commitment to OH&S and successfully achieve ISO 45001 certification. Preparation is key to ensuring that the audit process is smooth, effective, and aligned with safety goals.

Author's Personal Note

During my professional work with organizations implementing ISO standards, I have often seen that many companies initially view the requirements of ISO 45001 as a collection of technical clauses and documentation tasks. However, the true value of these requirements lies far beyond compliance with a standard.

The structure of ISO 45001 is designed to guide organizations toward a disciplined and proactive management approach. Each requirement—from understanding the context of the organization to leadership involvement, risk planning, operational control, and continuous improvement—forms part of an integrated system that strengthens the entire organization.

In my experience, the most successful organizations are those that treat these requirements not as obligations but as management tools. When leadership actively supports the system, employees participate in safety practices, and processes are continuously monitored and improved, the Occupational Health and Safety Management System becomes a natural part of daily operations.

Small enterprises often have a unique advantage in implementing these requirements. Their organizational structures are usually less complex, communication between management and employees is more direct, and decisions can be implemented quickly. With the right commitment, small organizations can build highly effective and responsive safety systems.

Over the years, I have seen companies transform their working environments simply by applying these principles consistently. What begins as an effort to comply with an international standard often evolves into a broader culture of responsibility, accountability, and continuous improvement.

Ultimately, the requirements of ISO 45001 should not be viewed as restrictions but as a framework that empowers organizations to protect their most valuable asset—their people—while building a stable and sustainable future.

— Dr. Mohamed-Ali Ibrahim

CHAPTER 5

Overcoming Challenges in Certification for Small Enterprises

Achieving **ISO 45001:2018** certification can be an intimidating process for small businesses due to the scale of the changes it demands in terms of both practices and resources. Although the benefits of certification—such as improved safety, compliance, and business growth—are well worth the effort, small enterprises often encounter several challenges along the way. This chapter explores some of the **common certification challenges** faced by small businesses and provides practical solutions to overcome them.

Common Certification Challenges for Small Businesses

1. Limited Resources (Financial, Human, and Time)

One of the most common challenges small businesses face when pursuing **ISO 45001 certification** is limited resources. Small businesses typically have fewer financial, human, and time resources, which can make it difficult to allocate enough attention and funding to the certification process.

Challenges:

- **Financial Constraints**: The cost of certification (including external audits, training, consulting fees, and necessary safety improvements) can be a barrier for small businesses with tight budgets.
- **Limited Staff**: Small businesses may lack the personnel necessary to manage both day-to-day operations and the extensive work required to implement and maintain the OH&S management system.
- **Time Constraints**: Small businesses often struggle to find the time to devote to the certification process, as employees are typically juggling multiple tasks, and operational priorities may compete with safety initiatives.

Solutions:

- **Outsource Expertise**: Small businesses can consider hiring external consultants or OH&S experts to assist with system development and audit preparation. Consultants can help small businesses save time by providing specialized knowledge and streamlining the process.
- **Phased Implementation**: Small businesses can adopt a phased approach to implementation. By gradually incorporating the necessary OH&S system elements over time, businesses can spread out costs and workloads, making the process more manageable.
- **Leverage Online Resources**: Online training programs, templates, and software can provide affordable, flexible options for employees to receive training and keep track of documentation. Additionally, digital tools can simplify record-keeping and reduce manual work.

2. Resistance to Change and Lack of Employee Engagement

Another common challenge is employee **resistance to change**. When small businesses implement new processes, such as those required for ISO 45001 certification, employees may feel uncertain or resistant, especially if they do not fully understand the benefits of the new system.

Challenges:

- **Employee Resistance**: Employees may be accustomed to established procedures and might be hesitant to adopt new safety protocols or engage in additional training.
- **Lack of Engagement**: Employees may not feel personally involved in the development or implementation of the OH&S management system, which could lead to a lack of participation in safety initiatives.

Solutions:

- **Involve Employees Early**: To overcome resistance, small businesses should actively involve employees in the process from the beginning. Engaging employees in discussions about the benefits of certification and seeking their input on safety practices will foster a sense of ownership and reduce resistance.

- **Provide Clear Communication**: Management should communicate the benefits of ISO 45001 certification in terms that employees can understand, highlighting how it will improve their own health and safety, as well as that of their colleagues.
- **Provide Training and Support**: Invest in employee training to increase awareness and ensure that staff members feel competent and comfortable with new safety procedures. A clear training plan helps employees understand the purpose behind the changes and boosts their confidence.

3. Lack of Clear Understanding of ISO 45001 Requirements

ISO 45001:2018 is a complex standard, and small businesses may lack the in-depth knowledge required to implement and maintain the OH&S management system. Understanding the full scope of ISO 45001 requirements, including risk assessments, legal compliance, and employee involvement, can be overwhelming for those unfamiliar with the standard.

Challenges:

- **Complexity of ISO 45001**: The detailed requirements of the standard may appear intimidating for small businesses without a dedicated OH&S team or safety manager.
- **Unclear Expectations**: Without clear guidance, small businesses may struggle to interpret the ISO 45001 requirements in a way that fits their operations.

Solutions:

- **Seek External Guidance**: Small businesses can benefit from external consultants, trainers, or certification bodies that specialize in ISO 45001. These experts can provide guidance on how to meet specific requirements and ensure the business is on the right track.
- **Simplify the Process**: Break down the implementation of the OH&S management system into manageable steps. Focus on one key element at a

time (e.g., policy creation, risk assessment, employee training) rather than attempting to tackle the entire system at once.

- **Leverage Online Resources**: ISO 45001 resources, including templates, guides, and checklists, are readily available online. Small businesses can use these to better understand the requirements and streamline the process.

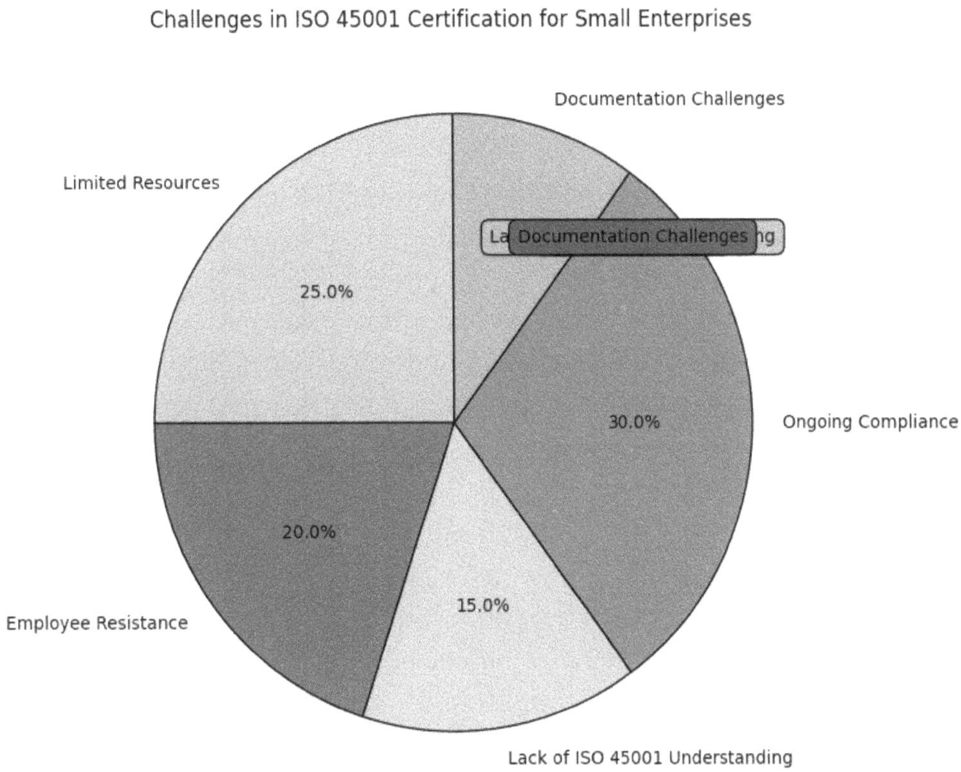

4. Managing Ongoing Compliance and Continuous Improvement

Achieving ISO 45001 certification is only the beginning. Maintaining ongoing compliance and continuously improving the OH&S management system can be a challenge for small businesses, especially when day-to-day operations demand attention and resources.

Challenges:

- **Continuous Improvement**: ISO 45001 emphasizes continuous improvement, requiring businesses to regularly review and update their safety management practices. Small businesses may find it challenging to prioritize this continuous review while managing other business activities.

- **Ongoing Monitoring**: Small businesses must regularly monitor and measure safety performance, conduct internal audits, and review risk assessments. Keeping up with these activities can become overwhelming if safety practices are not integrated into the day-to-day operations.

Solutions:

- **Establish Regular Review Cycles**: Implement a regular cycle of reviews, audits, and employee feedback sessions to ensure the OH&S system remains effective. Set aside specific times for management to evaluate safety practices and make adjustments as needed.
- **Integrate OH&S into Daily Operations**: Rather than treating OH&S as a separate task, integrate safety measures into everyday business activities. For example, regular safety briefings and daily checks can help maintain a continuous focus on health and safety.
- **Encourage Employee Feedback**: Continuously gather feedback from employees regarding the effectiveness of safety measures. Employee involvement helps identify areas that need improvement and fosters a culture of ongoing safety awareness.

5. Documentation and Record Keeping

ISO 45001 requires detailed documentation and record-keeping to ensure that the OH&S management system is functioning properly. For small businesses, maintaining accurate records can be time-consuming and may be seen as an additional burden.

Challenges:

- **Time-Consuming Documentation**: The process of documenting risk assessments, safety procedures, training records, and corrective actions can be overwhelming for small businesses with limited staff.
- **Lack of Centralized System**: Without an effective document management system, small businesses may struggle to maintain, organize, and retrieve the necessary records.

Solutions:

- **Implement Document Management Systems**: Use digital tools or software to streamline documentation and record-keeping. This makes it easier to store, update, and retrieve records when needed for audits.
- **Use Templates and Checklists**: Utilize ISO 45001 templates, checklists, and predefined formats for documentation. These tools simplify record-keeping and ensure consistency across the organization.
- **Designate Responsibility**: Assign responsibility for documentation and record-keeping to a designated person or team to ensure it is consistently maintained and up to date.

Conclusion: Overcoming Challenges for Successful Certification

While small businesses face numerous challenges when pursuing **ISO 45001:2018 certification**, understanding these challenges and implementing effective solutions is key to overcoming them. By addressing resource limitations, engaging employees, understanding ISO 45001 requirements, ensuring ongoing compliance, and streamlining documentation, small enterprises can successfully navigate the certification process.

With the right preparation, support, and commitment, small businesses can not only achieve certification but also build a robust OH&S management system that improves employee safety, enhances business performance, and strengthens their competitive position in the market.

Addressing Non-Conformities in Small Enterprises

Non-conformities refer to instances where a business fails to meet the required standards or practices set by ISO 45001:2018 or any internal safety protocols. Addressing these non-conformities is a critical aspect of the **ISO 45001 certification process**, as it helps ensure that the Occupational Health and Safety (OH&S) management system is continually improving and that the organization is compliant with both regulatory requirements and the ISO standard.

For **small enterprises**, addressing non-conformities can be challenging, but it is essential for ensuring the effectiveness of the OH&S system, reducing risks, and maintaining long-term compliance. Below, we will explore the process for identifying, addressing, and preventing non-conformities, along with strategies for sustaining compliance.

1. Understanding Non-Conformities

Non-conformities can be classified into two types:

A. Major Non-Conformities

- These are significant deviations from ISO 45001 requirements, which can compromise the effectiveness of the OH&S management system. If major non-conformities are found during an audit, they must be corrected before ISO 45001 certification can be granted or maintained.
- **Examples**: Lack of risk assessments, failure to provide adequate employee training, or failure to implement necessary safety controls.

B. Minor Non-Conformities

- These are less critical but still need to be addressed. While minor non-conformities do not immediately affect the integrity of the OH&S system, they should be rectified to prevent them from escalating into major issues.
- **Examples**: Inconsistent documentation of safety procedures or incomplete training records.

2. Identifying Non-Conformities

The first step in addressing non-conformities is identifying them. Non-conformities can arise from various sources, such as internal audits, employee feedback, incident reports, or external audits.

Key Sources for Identifying Non-Conformities:

- **Internal Audits**: Conduct regular internal audits to assess the performance of the OH&S system. These audits should focus on identifying areas where the organization is not meeting ISO 45001 standards.

- **Employee Feedback**: Employees who are directly involved in safety practices may notice issues that management overlooks. Encourage open communication and provide a formal mechanism for employees to report hazards or issues.
- **Incident Reports**: Review incident reports and accident investigations to identify recurring problems or gaps in the safety management system.
- **External Audits**: The external ISO 45001 auditor will provide feedback on any non-conformities discovered during the audit, which can then be used to make corrective actions.

3. Addressing Non-Conformities: Corrective and Preventive Actions

Once a non-conformity is identified, it's essential to take immediate action to correct it and implement preventive measures to avoid recurrence.

Key Steps in Addressing Non-Conformities:

A. Corrective Actions

Corrective actions are taken to address the root causes of non-conformities and prevent them from happening again. These actions should be well-documented, timely, and effective.

Steps for Corrective Actions:

1. **Root Cause Analysis**: Perform a thorough investigation to understand why the non-conformity occurred. Use tools such as **5 Whys** or **Fishbone Diagram** to determine the underlying cause.
2. **Develop a Corrective Action Plan**: Once the root cause is identified, develop an action plan to fix the issue. The plan should include clear steps, timelines, and responsible individuals.
3. **Implement the Corrective Action**: Take immediate action to resolve the non-conformity. This might involve revising safety procedures, providing additional training, or repairing faulty equipment.
4. **Monitor the Effectiveness of the Action**: Once the corrective action is implemented, track its effectiveness to ensure that the problem has been resolved and will not recur.

B. Preventive Actions

Preventive Actions are measures taken to eliminate the root cause of potential non-conformities before they occur.

Steps for Preventive Actions:

1. **Proactively Identify Risks**: Conduct regular risk assessments and look for areas where issues may arise in the future.
2. **Establish New Controls or Procedures**: Implement controls or processes that prevent similar problems from occurring. This could involve tightening up risk assessments, improving employee training, or upgrading equipment.
3. **Monitor for New Issues**: Continuously monitor for emerging risks or non-conformities to prevent them from affecting the OH&S system.

4. Documenting Non-Conformities and Actions Taken

Proper documentation is essential to demonstrate that corrective and Preventive Actions have been effectively implemented. This also helps in future audits, as auditors will expect to see clear records of actions taken to resolve non-conformities.

Key Documentation to Maintain:

- **Non-Conformity Reports**: A detailed report documenting the identified non-conformity, its cause, and the steps taken to correct it.
- **Corrective Action Records**: Documentation of the corrective actions, including who is responsible, deadlines, and the results of the actions taken.
- **Preventive Action Plans**: Records of Preventive Actions put in place to eliminate risks before they occur.
- **Follow-up Audits/Reviews**: Records of follow-up audits or reviews to verify that corrective actions were effective.

5. Ongoing Monitoring and Continuous Improvement

ISO 45001 places a strong emphasis on **continuous improvement**. After addressing non-conformities, small businesses must continue to monitor the effectiveness of the OH&S management system and make improvements as necessary.

Key Strategies for Ongoing Monitoring:

- **Regular Audits**: Perform regular internal audits to verify that corrective actions are working and to identify new potential non-conformities.
- **Safety Performance Metrics**: Track key safety performance indicators, such as incident rates, near misses, and employee participation in safety programs, to gauge the effectiveness of the system.
- **Employee Feedback**: Continuously engage employees to obtain feedback on safety practices and their perceptions of any new risks or areas for improvement.

Continuous Improvement:

- **Analyze Trends**: Look at trends over time to identify recurring safety issues. Continuous improvement involves adjusting practices based on this analysis to prevent future incidents.
- **Implement Best Practices**: Adopt best practices from within the organization or from other businesses to further enhance the OH&S management system.

6. Role of External Auditors in Non-Conformity Resolution

External auditors play an important role in identifying non-conformities during the certification process. They provide independent assessments and offer recommendations for improvement.

Key Responsibilities of External Auditors:

- **Identify Non-Conformities**: External auditors will assess the OH&S system against ISO 45001 requirements and identify areas where the system is not in compliance.
- **Recommend Corrective Actions**: If non-conformities are found, the auditor will provide recommendations for corrective actions. These suggestions help businesses address issues effectively and improve their safety systems.
- **Verify Corrective Actions**: The external auditor may revisit the business after corrective actions have been implemented to ensure that the actions were effective and that compliance is maintained.

7. Summary Table: Addressing Non-Conformities in Small Enterprises

Step	Description	Key Focus
Root Cause Analysis	Investigate the cause of the non-conformity using tools like the 5 Whys or Fishbone Diagram.	Identify the underlying issue causing the problem
Develop Corrective Action	Create a plan to address the root cause and prevent future occurrences.	Action plan, responsibility, and timeline
Implement Corrective Action	Take immediate steps to resolve the issue.	Ensure corrective actions are implemented properly
Monitor Effectiveness	Check if the corrective action effectively solved the issue.	Ensure the issue does not recur
Preventive Actions	Implement measures to prevent similar non-conformities from occurring.	Proactively identify and address potential issues
Documentation	Maintain records of non-conformities, corrective actions, and preventive measures.	Document all actions taken for future reference

Conclusion

Addressing **non-conformities** and implementing **corrective and Preventive Actions** is essential for maintaining a compliant and effective **ISO 45001:2018** OH&S management system. Small businesses must have a proactive approach to identifying, addressing, and preventing non-conformities, ensuring that the OH&S system is continuously improving. By documenting the actions taken and involving employees in safety efforts, small enterprises can build a culture of safety that leads to long-term success and compliance with ISO 45001.

Author's Personal Note

In every organization I have worked with, challenges during the certification process are inevitable. No management system is perfect from the beginning, and every organization—regardless of its size or industry—encounters obstacles when implementing new standards such as ISO 45001.

Over the course of my auditing and consulting experience, I have seen that the organizations which ultimately succeed are not those that avoid difficulties, but those that respond to them with openness and determination. Non-conformities, for example, are often misunderstood as failures. In reality, they are valuable indicators that reveal where a system can be strengthened.

A well-functioning Occupational Health and Safety Management System is built through continuous learning. Each corrective action, each improvement, and each lesson learned from a challenge contributes to a stronger and more resilient organization.

Small enterprises often show remarkable adaptability when facing these challenges. Their flexibility, close teamwork, and direct communication allow them to respond quickly and effectively to issues identified during audits or internal reviews. This agility can become a powerful advantage in maintaining and improving a safety management system.

Throughout my professional journey, I have observed that the most successful organizations treat certification not as a final destination but as the beginning of a continuous improvement process. The real achievement lies in building a workplace where safety, responsibility, and respect for employees become part of the organizational culture.

If there is one message I would like readers to take from this book, it is this: true excellence in occupational health and safety is achieved not through perfection, but through commitment, learning, and continuous improvement.

— Dr. Mohamed-Ali Ibrahim

Dear Reader,

Congratulations on taking the essential steps toward achieving **ISO 45001:2018 certification** for your small enterprise! By focusing on Occupational Health and Safety (OH&S), you are not only safeguarding your most valuable asset—your employees—but also positioning your business for long-term growth, compliance, and success.

The journey to certification may present challenges, but remember, the **ISO 45001** standard is designed to guide you toward continuous improvement and ensure that your workplace remains safe, healthy, and productive. As you move forward, it's important to stay committed to the process, keep your employees engaged, and maintain a proactive approach to health and safety.

Throughout this guide, we've explored the necessary steps, from understanding ISO 45001 requirements to addressing non-conformities, managing resources, and preparing for audits. It's essential that small businesses implement these systems systematically, with clear documentation, proper employee training, and regular reviews to ensure compliance. The benefits of ISO 45001—improved safety, legal compliance, reduced risks, and increased employee satisfaction—are well worth the effort.

Remember, achieving **ISO 45001** certification is not just about meeting regulatory requirements; it's about creating a culture of safety that fosters employee well-being and business resilience. Certification will enhance your organization's reputation, improve operational efficiency, and ultimately lead to better productivity and fewer incidents.

As you continue this journey, focus on **continuous improvement**. Stay open to feedback, regularly assess and improve your OH&S management system, and strive to maintain a work environment where safety and health are a top priority.

Kind Regards,

Dr. Mohamed-Ali Ibrahim
Austria, Vienna

PART 2

Establishing an Occupational Health and Safety Management System for Small Enterprises

A Practical Guide to ISO 45001:2018 Implementation for Small Enterprises Across All Business Sectors

CHAPTER 6

Building the Foundation of Your OH&S Management System in Small Enterprises

---

Steps for Developing Your OH&S Policy and OH&S Management System for Small Businesses

An Occupational Health and Safety Management System does not begin with a certificate hanging on a wall. It begins with structure. For small enterprises, that structure is the difference between a workplace that reacts to problems after they happen and a workplace that manages risks before they grow teeth. The foundation of the system is built through two closely linked elements: the **OH&S policy** and the **OH&S Management System**.

The **policy** sets the direction. It explains what the organization is committed to doing in relation to Occupational Health and Safety. The **OH&S Management System** turns that commitment into working practice. It defines how the system will function, who is responsible, how hazards will be identified, how risks will be controlled, how Workers will be involved, how legal duties will be tracked, and how improvement will be maintained.

For small businesses, this stage is critical because operations are often fast, people perform multiple roles, and safety can easily become informal. Informal systems may feel convenient, but they are unreliable. One day they seem fine; the next day they produce confusion, inconsistency, or an incident that nobody saw coming. A formal but practical OH&S foundation gives order, clarity, and accountability.

This chapter explains in full detail how small enterprises can build that foundation properly under ISO 45001:2018.

Understanding the Purpose of the OH&S Policy and OH&S Management System

Before writing anything, management should understand what these two elements are meant to achieve.

The **OH&S policy** is not a decorative statement. It is the official commitment of the enterprise to protect Workers and provide safe and healthy working conditions. It expresses management's intentions and direction. It tells people inside and outside the business what the organization stands for in relation to health and safety.

The **OH&S Management System** is the practical structure that supports the policy. It includes responsibilities, consultation arrangements, **risk assessment** methods, operational controls, training arrangements, emergency planning, communication methods, document control, monitoring, and review. If the policy says, 'we are committed,' the **management system** answers, 'this is how.' The **management system** not only identifies **risks** but also considers **opportunities** to improve the organization's health and safety performance.

A small enterprise needs both. **An OH&S Management System** without A policy without an OH&S Management System is like a promise without a plan. An OH&S Management System without a policy is like a machine with no steering.

Why Small Enterprises Need a Strong Foundation

Small businesses often work with limited resources, shared duties, and close operational pressure. That makes a strong system foundation even more important. In large organizations, responsibilities may be spread across departments. In small enterprises, one missing action or one unclear instruction can affect the whole business quickly.

A weak foundation creates several problems. Safety becomes dependent on memory. Workers receive different instructions from different people. Hazards are recognized only after incidents occur. Legal requirements are assumed rather than tracked. Corrective actions are delayed because nobody is clearly responsible. Training becomes irregular. Reporting becomes inconsistent. In other words, the system becomes a polite mess wearing a high-visibility vest.

A strong foundation helps the small business create order. It allows management to define what matters, assign responsibility, identify risks, control hazards, involve Workers, and monitor results. It also supports legal compliance, business continuity, worker confidence, and audit readiness.

I-Shape Diagram for Building the OH&S System Foundation

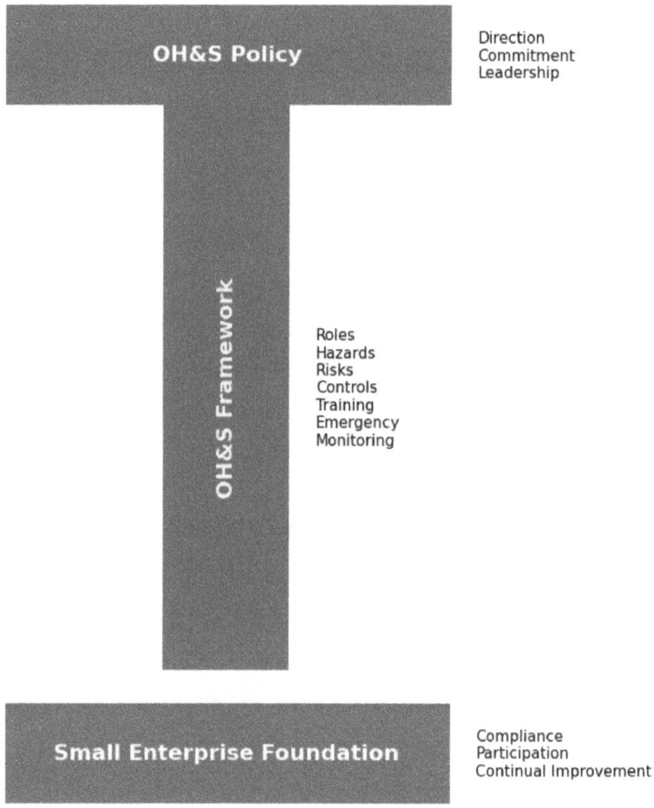

Step 1: Understand the Business and Its Activities

The first step in developing the OH&S policy and OH&S Management System is to understand the actual business. That sounds obvious, but this is where many businesses stumble. They begin by copying a generic safety policy from another organization without first examining their own operations, hazards, and working conditions.

Management should identify what the business does on a normal day. What products or services are provided? What tasks are performed? What machinery, tools, vehicles, or equipment are used? What substances are handled? Where does the work take place? Who carries out the work? Are contractors involved? Are temporary Workers used? Is work performed indoors, outdoors, at customer sites, in vehicles, in workshops, in offices, in kitchens, in clinics, or in mixed environments?

This operational understanding is essential because the policy and OH&S Management System must match reality. A small retail business faces different safety issues than a small warehouse. A maintenance company faces different risks than a school. A small fabrication unit, transport operator, office-based consultancy, food business, or medical practice all require an OH&S Management System, but the details of that system will differ.

Without understanding the business properly, the policy becomes generic and the OH&S Management System becomes weak.

Step 2: Review Internal and External Issues

ISO 45001:2018 expects the organization to understand the issues that affect its OH&S Management System. These include both internal and external factors.

Internal issues may include the size of the workforce, competence levels, communication style, management structure, shift work, maintenance arrangements, physical conditions, equipment age, supervision quality, worker behavior, and the overall safety culture. If the workplace is busy and noisy, that matters. If several Workers perform different jobs without formal supervision, that matters. If the business relies heavily on verbal instruction instead of documented procedures, that also matters.

External issues may include legal requirements, client expectations, contractor interfaces, insurance conditions, landlord obligations, regulatory inspections, environmental conditions, public access, neighboring businesses, and sector-specific risks. A small enterprise working in rented premises may depend on the landlord for some building safety issues. A contractor working on customer sites may face client-specific safety controls. A food business may have hygiene and chemical handling issues alongside workplace safety obligations.

Reviewing these issues helps management understand the environment in which the OH&S Management System will function. An OH&S Management System built without this understanding is likely to miss important risks or responsibilities.

Step 3: Identify Workers and Other Interested Parties

The next step is to determine who is affected by the OH&S Management System and whose needs and expectations matter. In ISO language, these are the relevant interested parties.

The most important group is the Workers. In small enterprises, this may include permanent **Workers**, part-time Workers, temporary Workers, apprentices, supervisors, drivers, technicians, office staff, cleaners, and in some cases owner-managers who perform operational work. Workers need safe equipment, clear procedures, proper instruction, suitable protective equipment where required, emergency support, and confidence that hazards will be addressed.

Other interested parties may include contractors, subcontractors, visitors, customers, suppliers, landlords, insurers, emergency responders, and regulatory authorities. Each of these parties may influence or be influenced by the business's OH&S performance.

A small enterprise should consider what these parties need and expect. Workers may expect clear reporting channels and fair treatment. Regulators expect legal compliance. Contractors need coordination and communication. Emergency services need clear access and emergency information. Clients may require safe delivery of contracted work. These needs should inform the policy and the OH&S Management System.

Step 4: Define the Scope of the OH&S Management System

A strong OH&S foundation requires a clearly defined scope. The scope explains what parts of the business are covered by the system.

For a small enterprise, this should include the relevant locations, activities, services, processes, and persons under the organization's control or influence. The scope may apply to the whole business or to specific sites, but it should be factual and realistic. It should not hide risky activities behind vague wording.

For example, if the business operates a workshop, warehouse, and office, the scope should make that clear. If the business performs off-site installation work or maintenance work at client premises, that should also be included where relevant. If contractors work on behalf of the organization in activities that affect OH&S performance, their role should be considered in the scope and OH&S Management System.

A weak or unclear scope causes confusion. People begin asking where the system applies, who is covered, and what activities are actually controlled. That confusion weakens the entire management system.

Step 5: Draft the OH&S Policy

Once the business context is understood, the OH&S policy can be prepared. The policy should be written clearly and should reflect the real nature of the enterprise.

A proper OH&S policy for a small business should include commitment to:

- providing safe and healthy working conditions
- preventing work-related injury and ill health
- fulfilling applicable legal and other requirements
- eliminating hazards and reducing OH&S risks
- consulting Workers and supporting their participation
- continually improving the OH&S Management System

These commitments are central. They are not random nice-sounding phrases. They represent the essential direction of the system under ISO 45001:2018.

The policy should also be suitable to the size, purpose, context, and risk profile of the business. A small logistics operator may need stronger emphasis on vehicle safety, manual handling, and site coordination. A small office-based business may focus more on ergonomics, stress, emergency arrangements, and contractor controls. A one-size-fits-all policy usually fits nobody particularly well.

The policy should be concise enough to communicate clearly, but meaningful enough to guide action.

Step 6: Secure Leadership Commitment

The policy must be approved and supported by top management. In a small enterprise, leadership commitment is especially visible because management is usually close to daily operations.

Top management may consist of the owner, managing director, general manager, or a small leadership team. These individuals must not only sign the OH&S policy but also show commitment through action. They should provide resources, support hazard reporting, require compliance with procedures, review OH&S performance, and respond properly to incidents and risks.

This is where many systems either gain credibility or collapse into performance art. If management says safety is important but tolerates unsafe shortcuts, Workers will quickly understand that the policy is ceremonial. A signed policy cannot compensate for careless management behavior.

Leadership commitment should therefore be visible in practical decisions, not only in documents.

Step 7: Build the OH&S Management System Around the Policy

After the policy is established, the organization must build the OH&S Management System that supports it. This is the practical operating structure of the OH&S Management System.

The OH&S Management System should define how the organization will:

- identify hazards
- assess OH&S risks
- determine legal requirements
- assign responsibilities
- consult and involve Workers

- provide training and awareness
- establish operational controls
- prepare for emergencies
- report and investigate incidents
- maintain documented information
- monitor performance
- review the system
- improve the system over time

For small enterprises, the OH&S Management System should be simple, workable, and clearly connected to operations. It should not be overdesigned. A bulky OH&S Management System that nobody uses is no better than no OH&S Management System at all. On the other hand, a vague OH&S Management System that says little more than "we will try our best" is too weak to provide control.

The best OH&S Management System for a small business is one that people can apply in normal working conditions without confusion.

Step 8: Define Roles and Responsibilities

Every OH&S Management System needs clarity about who is responsible for what. In small enterprises, this is especially important because one person may perform several roles.

The **OH&S Management System** should define responsibilities for policy approval, **risk assessment**, workplace inspections, training coordination, incident reporting, corrective action follow-up, emergency response, legal compliance monitoring, and management review. Even if one person performs several of these duties, the responsibilities must still be clearly identified.

Workers should know their own responsibilities too. These may include following procedures, using protective equipment correctly, reporting hazards, attending training, cooperating with investigations, and participating in consultation activities.

Without clear responsibilities, the system becomes dependent on assumption. And assumption is a mischievous creature; it always shows up smiling right before things go wrong.

Step 9: Establish a Based on based on risk assessment methods defined by the organization in line with ISO 45001:2018 Clause 6.1.2" to align with risk-based thinking. Process

Based on based on risk assessment methods defined by the organization in line with ISO 45001:2018 Clause 6.1.2" to align with risk-based thinking. is a core part of the OH&S Management System. The business must have a structured way to identify what can cause injury or ill health.

This should include hazards arising from routine work, non-routine work, maintenance, cleaning, movement of people and vehicles, equipment use, manual handling, chemicals, noise, electricity, ergonomic strain, contractor activities, emergency situations, and workplace changes.

Based on based on risk assessment methods defined by the organization in line with ISO 45001:2018 Clause 6.1.2" to align with risk-based thinking. should consider how work is actually done, not just how procedures say it is done. Workers often adapt tasks based on pressure, space limitations, time, weather, or available tools. A realistic system examines actual practice.

The process should also consider future changes. New equipment, new substances, layout changes, new clients, temporary projects, or new work patterns may introduce new hazards. Based on based on risk assessment methods defined by the organization in line with ISO 45001:2018 Clause 6.1.2" to align with risk-based thinking. should therefore be ongoing, not one-time.

Step 10: Establish a Risk Assessment Method

After hazards are identified, the business must assess the **risks**. This means deciding how serious the consequences could be, how likely the event is, who may be exposed, and whether current controls are adequate.

The method does not need to be overly complicated. A small enterprise can use a practical **risk assessment** matrix or another simple scoring method, as long as it is consistent and effective. What matters is not complexity; what matters is structured judgment.

Risk assessment should help the organization decide:

- which **hazards** require immediate action
- which controls are missing or weak
- which **risks** can be tolerated temporarily
- which **risks** require elimination, substitution, engineering controls, administrative controls, or PPE

A **risk assessment** should always lead to control decisions. If the organization assesses **risk** but fails to act, the process becomes an expensive way to admire danger on paper.

Step 11: Determine Legal and Other Requirements

A small enterprise must know what legal and other OH&S requirements apply to its operations. Compliance cannot be based on memory, assumption, or the ancient managerial tradition of "I'm sure we're probably fine."

The OH&S Management System should define how the business identifies relevant legal requirements and other obligations. These may include workplace safety laws, fire safety duties, first aid requirements, machinery safety rules, chemical handling regulations, welfare provisions, working hours obligations, contractor management duties, vehicle safety rules, and sector-specific standards.

The organization should also consider other requirements it chooses to follow, such as client rules, insurer expectations, lease conditions, or industry codes. Once these requirements are identified, they should be integrated into the OH&S Management System.

The OH&S Management System should also explain how the organization keeps this information updated and how compliance is checked.

Step 12: Establish Worker Consultation and Participation

Worker consultation and participation are essential under ISO 45001. Workers are often the first to notice unsafe conditions, weak controls, impractical procedures, or operational drift. A business that fails to involve Workers is cutting itself off from one of the most useful sources of OH&S information.

The OH&S Management System should explain how Workers are informed, consulted, and involved. This may include safety meetings, toolbox talks, direct reporting channels, participation in risk assessments, involvement in incident investigations, and contribution to corrective actions.

Consultation should be active, not symbolic. Workers should be able to raise issues and receive feedback. If they report hazards and hear nothing back, trust erodes, and future reporting declines. The organization should create an environment where participation is encouraged and where barriers to participation are removed as much as possible.

In small enterprises, this can be highly effective because communication lines are often shorter and more direct.

Step 13: Define Operational Controls

The OH&S Management System must explain how identified risks will be controlled during operations. These are the actual rules, methods, and measures that reduce exposure to harm.

Operational controls may include safe work procedures, machine guarding, housekeeping rules, maintenance schedules, contractor controls, manual handling methods, storage arrangements, electrical isolation procedures, supervision rules, access restrictions, signage, inspection routines, and use of personal protective equipment.

The organization should apply the hierarchy of controls where possible. Elimination is the strongest option. Substitution comes next. Engineering controls and administrative controls follow. PPE should be used where needed, but it should not be the first answer to every risk.

A small business that responds to every hazard with 'wear gloves and be careful that approach may have charm in a folk-medicine kind of way, but it is not enough.

Step 14: Establish Competence, Training, and Awareness

A policy and OH&S Management System are only effective if Workers understand what is expected and are competent to perform their duties safely.

The business should determine what competence is necessary for each role. This includes technical ability, understanding of hazards, knowledge of procedures, awareness of emergency response, and ability to use equipment and protective measures correctly.

Training should include induction for new Workers, task-specific instruction, refresher training where necessary, and additional training when processes, equipment, or risks change. Records should be maintained to show what training has been provided and to whom.

Awareness is also important. Workers should understand the OH&S policy, know the main hazards relevant to their work, understand why controls matter, and know how to report incidents, hazards, and unsafe conditions.

A system without competence is like handing someone a map written in a language they cannot read and then acting surprised when they get lost.

Step 15: Establish Communication and Documented Information Controls

The OH&S Management System should define how information is communicated internally and externally. Workers need clear instructions, updates, alerts, emergency information, and consultation opportunities. Contractors may need safety rules and hazard briefings. Emergency services may need specific access information. Regulators may require evidence of compliance.

The organization should also control documented information. This includes the policy, procedures, risk assessments, inspection forms, incident reports, training records, legal registers, maintenance records, meeting notes, and corrective action logs.

For small enterprises, the documentation should be sufficient to operate the system and show evidence of control. It should be current, accessible, and protected from uncontrolled changes. It should not be excessive or chaotic. Too little documentation weakens evidence. Too much documentation creates confusion and is often ignored.

The goal is controlled simplicity, not paper inflation.

Step 16: Include Emergency Preparedness and Response

Emergency preparedness should be built into the OH&S Management System from the start. The organization should identify foreseeable emergency situations relevant to its activities. These may include fire, electrical incidents, chemical spills, medical emergencies, gas leaks, equipment failures, vehicle incidents, or evacuation events.

The OH&S Management System should define emergency responsibilities, alarm arrangements, contact methods, first aid support, evacuation procedures, assembly points, emergency equipment, and where relevant, coordination with external responders.

Emergency arrangements should be communicated clearly to Workers. Training or drills may be required depending on the nature of the risks. Emergency preparedness should also be reviewed after drills, actual incidents, or major changes.

An emergency plan that exists only because somebody downloaded a template three years ago and never looked at it again is not preparedness. It is literary fiction.

Step 17: Establish Monitoring, Review, and Improvement Processes

A system must be checked if it is going to remain effective. The OH&S Management System should define what will be monitored, how often, by whom, and with what records.

Monitoring may include inspections, observations, incident trends, training completion, maintenance performance, corrective action status, legal compliance checks, and worker feedback. The business should decide what indicators matter most based on its risks and operations.

Management should also review the OH&S Management System periodically. This review should examine whether the system remains suitable, adequate, and effective. It should consider incidents, audit Observations and recommendations must be documented; implementation is per organization discretion., corrective actions, worker consultation results, legal changes, resource needs, and opportunities for improvement.

Continual improvement should be built into the system. When issues are found, the business should respond. When hazards change, controls should change. When performance is weak, action should follow. A system that is never reviewed will quietly drift until it becomes outdated and unreliable.

Step 18: Communicate the Policy and Keep It Relevant

Once the policy and OH&S Management System are established, the policy should be communicated throughout the organization. Workers should know that it exists and understand what it means in practice. It may be displayed in the workplace, included in induction, discussed during meetings, and referenced in training.

The policy should also be reviewed periodically. If the organization changes its activities, equipment, risk profile, legal obligations, or structure, the policy may need revision. A policy should remain relevant to the business, not preserved like a museum exhibit from a past era.

Relevance matters. If Workers see a policy that clearly reflects current operations, they are more likely to take it seriously. If they see a generic document with no practical connection to their work, they will treat it as background wallpaper.

Conclusion

Building the foundation of an OH&S Management System in a small enterprise requires more than producing a policy statement. It requires understanding the business, reviewing internal and external issues, identifying Workers and interested parties, defining the scope, drafting a suitable policy, securing management commitment, and creating a practical OH&S Management System that supports real control.

That OH&S Management System must define responsibilities, based on based on risk assessment methods defined by the organization in line with ISO 45001:2018 Clause 6.1.2" to align with risk-based thinking., risk assessment, legal compliance, worker consultation, operational controls, competence, documentation, emergency preparedness, monitoring, and improvement. When these elements are developed properly, the organization creates a working system rather than a ceremonial one.

For small businesses across all sectors, the objective is not complexity for its own sake. The objective is a system that is clear, practical, suitable, and strong enough to protect people while supporting compliance and continual improvement. A strong foundation makes the rest of the OH&S Management System possible. A weak one makes everything that follows wobble.

And wobbling is an entertaining feature in circus acts, but not in Occupational Health and Safety Management Systems.

Aligning OH&S Practices with Business Operations in Small Enterprises

For a small enterprise, Occupational Health and Safety cannot sit in a separate folder, handled only when an auditor visits or when an incident happens. If OH&S is treated as a side activity, it will remain weak, inconsistent, and reactive. To be effective, OH&S practices must be aligned with normal business operations. That means health and safety requirements should be built into the way the business plans work, assigns tasks, uses equipment, manages people, deals with contractors, purchases materials, and reviews performance.

In small businesses, this alignment is especially important because operations are usually lean, roles overlap, and decisions are made quickly. A single supervisor may handle production, staffing, maintenance coordination, and customer issues on the same day. If safety is not built into those decisions, it gets pushed aside by time pressure, cost pressure, or convenience. The result is usually a workplace that depends too much on habit and not enough on control.

Aligning OH&S with business operations means making safety part of how the business functions, not something added afterward. When a job is planned, the associated risks should be considered at the same time. When equipment is purchased, safety requirements should be part of the purchase decision. When staff are assigned, competence and training should be checked before work starts. When changes are introduced, their health and safety impact should be reviewed before implementation. In this way, safety becomes part of operational management rather than a disconnected activity.

A practical alignment starts with understanding the main business processes. Small enterprises should identify how work flows from one stage to another.

This may include sales, purchasing, delivery, storage, production, maintenance, administration, customer service, cleaning, waste handling, and contractor coordination. Each of these activities can create OH&S risks. If the business understands its process flow clearly, it becomes easier to identify where health and safety controls need to be integrated.

For example, in purchasing, the business should consider whether machinery is safe, whether chemicals are properly labeled, whether suppliers provide safety information, and whether personal protective equipment is suitable. In staffing, the business should consider whether Workers are competent, fit for the task, and properly instructed. In maintenance, the business should ensure isolation procedures, access controls, and repair arrangements are clear. In daily supervision, the business should include inspections, hazard reporting, and immediate correction of unsafe practices. These are not separate from operations; they are operations managed properly.

Another important aspect of alignment is planning. Daily work planning should include OH&S considerations. If the business plans overtime, it should consider fatigue. If it changes layout, it should consider movement routes, emergency exits, and manual handling implications. If it increases production speed, it should consider whether the faster pace creates additional risks. Small businesses often change conditions quickly, which makes it even more important to ensure that OH&S remains linked to operational decisions.

Training and communication also need alignment. Workers should not receive safety training as a one-off event that is forgotten the next day. Instead, training should match operational needs. A forklift operator needs practical instruction linked to the workplace layout and traffic conditions. A cleaner handling chemicals needs instruction relevant to those substances and work methods. A technician working off-site needs training relevant to customer environments and travel-related risks. This is where alignment becomes real: training matches work, and work reflects the control measures required by the OH&S Management System.

Monitoring must also be connected to business performance. Small enterprises often track delivery targets, customer deadlines, product quality, and cost control, but they may not track inspections, near misses, training completion, corrective actions, or incident trends with the same seriousness. A well-aligned OH&S Management System makes these safety indicators part of management review, not background noise. When safety performance is reviewed alongside operational performance, management sends a clear signal that health and safety is part of business control.

Alignment also means that changes in business activity should automatically trigger OH&S review. New equipment, new Workers, new products, new clients, new materials, new locations, and new schedules all affect risk. A small enterprise should create a habit of asking: what does this change mean for health and safety? That question should be part of normal operational management.

When OH&S is truly aligned with operations, several benefits appear. Hazards are identified earlier. Controls are applied more consistently. Workers receive clearer instructions. Management decisions become more balanced. Incidents reduce. Compliance improves. The OH&S Management System becomes stronger because it is used in practice rather than preserved as a formal requirement.

In short, aligning OH&S practices with business operations means that health and safety becomes part of how the enterprise functions every day. It moves from being a reaction to being a management discipline.

Defining Roles and Responsibilities for OH&S in Small Enterprises

A small enterprise cannot operate an effective OH&S Management System unless people know exactly what is expected of them. One of the most common weaknesses in small businesses is unclear responsibility. Safety tasks are assumed, shared loosely, or handled only when someone remembers. That approach leads to missed inspections, incomplete training, delayed corrective actions, weak incident reporting, and confusion during emergencies.

Defining roles and responsibilities for OH&S means assigning clear duties, authority, and accountability to the people involved in the system. This does not mean creating unnecessary bureaucracy. In a small enterprise, one person may perform several roles, and that is perfectly normal. What matters is not the number of job titles; what matters is clarity.

The starting point is top management. In a small business, this may be the owner, managing director, or general manager. Top management holds overall responsibility for the OH&S Management System. This includes approving the OH&S policy, providing resources, ensuring responsibilities are assigned, supporting worker participation, reviewing performance, and promoting continual improvement. Top management may delegate tasks, but it cannot delegate ultimate accountability for the system.

The next level often includes supervisors or operational managers. Their role is usually more direct and practical. They are responsible for implementing OH&S controls in daily work, monitoring safe behavior, ensuring Workers follow procedures, reporting incidents, stopping unsafe work when necessary, and communicating with Workers about hazards and controls. In small enterprises, supervisors are often the bridge between policy and practice. If they are not engaged, the system weakens quickly.

Workers also have defined responsibilities. They are responsible for following established procedures, using tools and protective equipment properly, reporting hazards and incidents, participating in training, cooperating with investigations, and contributing to consultation processes. Worker responsibility should never be confused with management responsibility. Workers must act safely, but management must provide the conditions, resources, and supervision needed to make safe work possible.

If the business has a designated safety coordinator or OH&S representative, that person may support documentation, inspections, legal registers, training records, and corrective action follow-up. However, this role should support the system, not carry the whole system alone like an exhausted mule dragged through a paperwork swamp. OH&S must remain integrated into management and operations, not isolated in one person's desk.

Contractors and external providers may also have OH&S responsibilities where relevant. If they work on-site or on behalf of the organization, the business should define what safety rules apply to them, who supervises them, how risks are communicated, and what responsibilities they hold for reporting hazards or incidents. A weak contractor-control arrangement can damage an otherwise decent OH&S Management System very quickly.

Emergency roles must also be clear. Who raises the alarm? Who coordinates evacuation? Who provides first aid? Who contacts emergency services? Who checks attendance at the assembly point? In a small enterprise, these roles should be simple and realistic. If emergency duties are unclear, confusion will appear exactly when the business can least afford it.

Responsibilities should be documented in a practical way. This may be done through job descriptions, role matrices, procedures, induction materials, or responsibility charts. The method is less important than the result: people should understand their duties clearly. The business should also communicate these responsibilities and review them whenever organizational changes occur.

Clarity in roles improves the system in several ways. It reduces duplication. It closes gaps. It strengthens accountability. It improves communication. It ensures safety tasks are not forgotten. It also makes audits easier because the organization can show who is responsible for inspections, risk assessments, training, incident reporting, corrective action, and review.

Most importantly, clearly defined roles help the business move from informal safety habits to controlled OH&S management. When people know what they are responsible for, safety becomes more reliable.

Table: Aligning OH&S Practices with Business Operations and Defining Responsibilities

Business Area / Function	How OH&S Should Be Aligned	Main Responsible Role	Typical OH&S Responsibilities
Top Management	OH&S included in business decisions, objectives, resourcing, and review	Owner / Director / General Manager	Approve policy, provide resources, assign responsibilities, review system performance, support continual improvement
Daily Operations	Safety controls built into work planning and supervision	Supervisor / Operations Manager	Ensure safe work practices, monitor compliance, correct unsafe acts, report incidents, communicate hazards
Purchasing and Procurement	Safety requirements included in buying decisions	Manager / Purchasing Officer	Purchase safe equipment, obtain safety data, ensure PPE suitability, verify contractor or supplier safety requirements
Staffing and Competence	Worker assignment based on skill, training, and fitness for task	Manager / HR / Supervisor	Confirm competence, arrange induction, assign training, verify qualifications, monitor capability
Maintenance and Equipment Control	Safety integrated into maintenance planning and repair activity	Maintenance Lead / Supervisor	Arrange inspections, ensure machine safety, support isolation/lockout where required, control repair risks
Risk Assessment	Risks reviewed before and during operations	Supervisor / OH&S Coordinator / Manager	Identify hazards, assess risks, recommend controls, review changes in conditions or tasks
Worker Participation	Worker input built into daily management and improvement	Supervisor / Manager	Encourage hazard reporting, hold briefings, consult Workers, involve them in investigations and assessments

Training and Awareness	Training linked directly to real tasks and operational risk	Supervisor / Trainer / OH&S Coordinator	Deliver induction, task-specific training, refresher training, maintain records, build awareness
Incident and Near-Miss Reporting	Reporting process integrated into daily business control	All Workers, Supervisor, Manager	Report incidents, investigate causes, implement corrective action, track completion
Emergency Preparedness	Emergency arrangements linked to workplace realities	Manager / Emergency Coordinator / First Aider	Raise alarm, guide evacuation, provide first aid, contact emergency services, review emergency response
Contractor Management	Contractors included in operational safety controls	Manager / Site Supervisor	Communicate site rules, review contractor risks, monitor contractor compliance, coordinate shared hazards
Performance Monitoring	Safety performance reviewed alongside operational performance	Top Management / Manager / OH&S Coordinator	Track inspections, corrective actions, training status, incidents, near misses, and review Observations and recommendations must be documented; implementation is per organization discretion.

Key Practical Point for Small Enterprises

In a small enterprise, the same person may appear in several rows of that table. That is normal. The important issue is not whether duties are spread widely or concentrated narrowly. The important issue is that duties are clearly defined, communicated, and carried out.

If OH&S responsibilities are vague, the system becomes fragile. If OH&S practices are disconnected from business operations, the system becomes artificial. But when responsibilities are clear and OH&S is built into operations, the management system becomes practical, reliable, and much easier to sustain over time.

Author's Personal Note

The development of an Occupational Health and Safety Management System is often viewed by small enterprises as a demanding task, particularly when resources, time, and personnel are limited. Yet the real strength of such a system does not lie in the size of the organization or the complexity of its documentation. Its strength lies in the commitment to protect people, to manage risks responsibly, and to build a working environment where safety is treated as a fundamental part of daily operations.

Small businesses form the backbone of most economies. They create employment, support communities, and drive innovation across countless sectors. Because of their scale, however, they often operate close to the ground, where the realities of work are immediate and the consequences of safety failures can be deeply personal. In such environments, even simple improvements in safety management can make a significant difference to the wellbeing of Workers and the stability of the business itself.

The ideas presented in this chapter are not intended to complicate operations or burden organizations with unnecessary procedures. Instead, they aim to encourage a structured and thoughtful approach to managing workplace health and safety. When safety becomes integrated into everyday decisions, when responsibilities are clearly understood, and when risks are managed with awareness and discipline, a small enterprise builds more than compliance—it builds trust, resilience, and long-term sustainability.

Every organization begins this journey from a different starting point. Some may already have strong informal safety practices, while others may be taking their first structured steps toward an Occupational Health and Safety Management System. What matters most is the willingness to improve, to learn from experience, and to treat the protection of Workers as an essential responsibility of leadership.

The hope is that the principles discussed throughout this chapter contribute to a clearer understanding of how small enterprises can strengthen their OH&S practices while maintaining operational flexibility. When safety becomes part of the culture of the organization, it supports not only compliance with ISO 45001 but also the creation of a workplace where people can perform their work with confidence and security.

Kind regards,
Dr. Mohamed-Ali Ibrahim
Vienna, Austria

CHAPTER 7

Risk Assessment and Based on based on risk assessment methods defined by the organization in line with ISO 45001:2018 Clause 6.1.2" to align with risk-based thinking. in Small Enterprises

---

Importance of Risk Assessment in Small Enterprises

Risk assessment is one of the most important parts of an Occupational Health and Safety Management System. For small enterprises, it is not just a formal requirement under ISO 45001:2018. It is a practical method for understanding where harm may occur, how serious that harm could be, who might be affected, and what controls are needed to reduce the danger. In simple terms, risk assessment helps the business see problems before those problems turn into injuries, illness, damage, downtime, or legal trouble.

In many small businesses, work moves quickly and responsibilities overlap. One employee may perform several different tasks in the same day. A supervisor may manage production, people, equipment, and customer deadlines all at once. Because of this, hazards can easily be missed if the business relies only on experience or routine habit. Risk assessment brings structure into that environment. It replaces guesswork with a method.

What Risk Assessment Means in Practical Terms

Risk assessment means looking carefully at work activities and asking a set of disciplined questions:

- What hazard is present?
- What kind of harm could happen?
- Who could be affected?
- How likely is the harm to occur?
- How severe could the consequences be?

- What controls already exist?
- Are those controls enough?
- What more needs to be done?

These questions sound simple, but they are powerful. They force the business to move beyond vague statements such as "this area is risky" or "Workers should be careful." Those kinds of statements may sound serious, but they do not control anything. Risk assessment turns concern into action.

Why Risk Assessment Is Especially Important for Small Enterprises

For a small enterprise, even one incident can have a major effect. In a large organization, a single injury may be absorbed more easily because staffing, finances, and systems are broader. In a small enterprise, the same event can hit much harder.

A serious accident may lead to:

- worker absence
- reduced productivity
- delayed customer commitments
- medical and compensation costs
- equipment damage
- legal and regulatory pressure
- stress on supervisors and management
- reputational harm

This is why risk assessment matters so much in small enterprises. It helps prevent the kind of disruption that a small business may find difficult to absorb.

Risk Assessment Supports Prevention Rather Than Reaction

One of the strongest reasons for carrying out risk assessment is that it shifts the business from a reactive mindset to a preventive one. A reactive business waits until something goes wrong and then responds. A preventive business tries to identify the danger before the incident happens.

This difference is enormous.

Without risk assessment, the business often learns through injury, complaint, equipment damage, or near misses. With risk assessment, the business learns by examining work systematically before harm occurs. That is a far more civilized method, and usually a much less expensive one too.

Shape Diagram: Importance of Risk Assessment in Small Enterprises

For example, risk assessment may reveal that:

- Workers are lifting loads in an unsafe way
- machine guards are missing or weak
- cleaning chemicals are stored poorly
- pedestrian and vehicle routes overlap
- staff are undertrained for a certain task
- electrical cables create trip or shock hazards
- emergency exits are blocked by stock or tools

If these issues are identified early, the business can correct them before they produce real harm.

Risk Assessment Helps Make Hidden Hazards Visible

Many workplace hazards become "normal" simply because people see them every day. Workers walk past them, work around them, and gradually stop noticing them. This is common in small businesses where routines become familiar and people assume that because nothing serious has happened yet, everything must be under control.

Risk assessment interrupts that false comfort.

It forces the organization to look again at routine activities, including:

- daily operations
- non-routine tasks
- maintenance work
- cleaning activities
- loading and unloading
- contractor work
- emergency situations
- changes in layout, equipment, or staffing

This fresh look often reveals hazards that had become invisible through familiarity.

Risk Assessment Strengthens Legal Compliance

Small enterprises have legal duties for health and safety, just like larger organizations. The law generally does not reduce responsibility simply because the business is small. Risk assessment helps show that the organization is taking reasonable and structured steps to identify and control hazards.

This is important because regulators, insurers, auditors, and clients may all want evidence that risks are being managed properly. A small enterprise that cannot show how it assessed its major risks is standing on shaky ground.

Risk assessment supports legal compliance by helping the business:

- identify workplace hazards
- evaluate exposure and consequences
- apply suitable control measures
- review whether controls remain effective
- document decisions and actions taken

In short, risk assessment helps demonstrate that the business is not ignoring its duties.

Risk Assessment Improves Operational Planning

Risk assessment is not separate from business operations. It should influence how work is planned and managed every day. When used properly, it becomes a management tool.

It can affect decisions about:

- task assignment
- staffing levels
- work scheduling
- machine use
- maintenance timing
- storage layout
- material handling
- contractor supervision
- emergency arrangements
- training priorities

For example, if a risk assessment shows that fatigue is becoming a concern during long shifts, management may need to review working hours or breaks. If it shows repeated manual handling strain, the business may need to change storage levels, reduce load weights, or introduce lifting aids. If it shows repeated slip hazards, then housekeeping, drainage, flooring, or cleaning methods may need review.

This is where risk assessment becomes useful in the real world. It helps management make better decisions instead of relying on assumption.

Risk Assessment Supports Consultation and participation of Workers

In small enterprises, Workers often know the job better than anyone reading a procedure at a desk. They know where shortcuts are taken, where equipment is unreliable, and where real pressure appears during the day. This makes worker participation extremely valuable in risk assessment.

When Workers are involved, the process becomes more accurate because it reflects actual work rather than idealized work.

Consultation and participation of Workers helps the business:

- identify hazards that management may miss
- understand how tasks are really performed
- test whether controls are practical
- improve communication and trust
- strengthen reporting culture

A risk assessment done without worker input may look tidy but still miss important realities. A risk assessment done with proper worker participation is usually much stronger.

Risk Assessment Helps Prioritize Resources

Small enterprises do not usually have unlimited money, time, staff, or equipment. That means they must use resources carefully. Risk assessment helps by showing which issues are most serious and which controls need priority.

Not every hazard carries the same level of risk. Some require urgent action. Others may already be adequately controlled. A structured assessment helps management decide where effort should go first.

This is one of the biggest practical benefits of risk assessment. It allows the business to act proportionately instead of either:

- overreacting to minor issues, or
- underreacting to major ones

That balance matters, especially in smaller organizations where resources must be directed intelligently.

Risk Assessment Supports Continual Improvement

Risk assessment is not a one-time exercise. Work changes. Staff change. Equipment changes. Layouts change. Production levels change. Customer demands change. New hazards appear, and old assumptions become outdated.

Because of this, risk assessment should be reviewed regularly and whenever significant changes occur.

This helps the business:

- keep controls relevant
- respond to new risks
- evaluate whether old controls still work
- improve the OH&S Management System over time

A small enterprise that never reviews its risks eventually ends up controlling yesterday's problems while today's hazards walk around freely wearing tomorrow's incident report.

Typical Hazard Areas in Small Enterprises

The exact hazards will vary by sector, but many small enterprises commonly face risks such as:

- slips, trips, and falls
- manual handling injuries
- machine and equipment hazards

- vehicle movement
- electrical hazards
- fire risks
- chemical exposure
- noise and vibration
- poor housekeeping
- ergonomic strain
- fatigue and workload pressure
- contractor-related risks
- emergency response weaknesses

The importance of risk assessment lies in identifying which of these hazards apply to the business and how serious they are in the specific workplace.

How Risk Assessment Strengthens the Whole OH&S Management System

Risk assessment is not an isolated activity. It influences many other parts of the OH&S Management System. If risk assessment is weak, the rest of the system often becomes weak as well.

A strong risk assessment process supports:

- relevant OH&S objectives
- effective operational controls
- proper training and awareness
- suitable emergency preparedness
- better monitoring and review
- meaningful audits
- stronger corrective actions

This is why auditors pay close attention to based on based on risk assessment methods defined by the organization in line with ISO 45001:2018 Clause 6.1.2” to align with risk-based thinking. and risk assessment. They are not looking for paperwork alone. They are looking for evidence that the organization understands its real risks and is controlling them in a sensible, structured way.

Key Practical Benefits of Risk Assessment

A well-performed risk assessment helps a small enterprise to:

- prevent injuries and ill health
- reduce downtime and disruption
- improve legal compliance
- prioritize actions and resources
- support worker participation
- strengthen management decisions
- improve business continuity
- support certification readiness under ISO 45001:2018

That is quite a lot of benefit from one disciplined process. Not bad for something often treated like a form-filling chore.

Conclusion

Risk assessment is important in small enterprises because it creates control where uncertainty would otherwise dominate. It helps the business identify hazards, evaluate their seriousness, protect Workers, support legal compliance, improve operational planning, and reduce the chance of incidents that could seriously affect a smaller organization.

For small businesses, risk assessment is not just a technical requirement. It is a survival-strengthening tool. It allows the enterprise to see clearly, act early, and manage work more responsibly. When carried out properly, it becomes one of the strongest practical foundations of the entire OH&S Management System.

Assessment and documented control measures are required to manage OH&S risks effectively Risk assessment is.

Techniques for Identifying Hazards in Small Businesses

Based on based on risk assessment methods defined by the organization in line with ISO 45001:2018 Clause 6.1.2" to align with risk-based thinking. is the starting point of effective Occupational Health and Safety management. A business cannot control what it has not identified, and it cannot improve what it does not understand. For small businesses, based on based on risk assessment methods defined by the organization in line with ISO 45001:2018 Clause 6.1.2" to align with risk-based thinking. is especially important because work is often fast-moving, resources are limited, and several activities may be carried out by the same people in the same space. In such conditions, hazards can easily hide inside routine work and become accepted as "normal."

A hazard is anything with the potential to cause injury, ill health, damage, or disruption. In a small enterprise, hazards may arise from machinery, manual handling, vehicles, chemicals, poor housekeeping, electricity, noise, repetitive work, stress, contractor activities, temperature, work at height, confined spaces, unsafe behavior, or badly planned work organization. Based on based on risk assessment methods defined by the organization in line with ISO 45001:2018 Clause 6.1.2" to align with risk-based thinking. is therefore the disciplined process of finding these sources of harm before they lead to incidents.

For small businesses, based on based on risk assessment methods defined by the organization in line with ISO 45001:2018 Clause 6.1.2" to align with risk-based thinking. should not be treated as a one-time exercise carried out only during setup or before an audit. It should be an ongoing activity linked to daily operations, supervision, planning, changes, incidents, and worker consultation. Hazards change when equipment changes, staffing changes, layouts change, production pressure increases, or new contractors enter the workplace. A based on based on risk assessment methods defined by the organization in line with ISO 45001:2018 Clause 6.1.2" to align with risk-based thinking. process must therefore be active, not sleepy.

Workplace Walk-Through Inspection

One of the most practical techniques for identifying hazards is the **Workplace Walk-Through Inspection**. This involves physically observing the work area, the equipment, the work methods, and the conditions in which tasks are performed. A walk-through helps identify visible hazards such as blocked exits, unsafe storage, damaged tools, poor lighting, housekeeping failures, slippery floors, unguarded machinery, exposed wiring, and congestion in movement routes.

In small businesses, this technique is particularly useful because the scale of operations often allows management and supervisors to directly inspect the areas where work takes place. The value of this technique increases when the inspection is structured rather than casual. A formal checklist can help ensure that important hazard categories are not overlooked.

Task Observation

Another effective technique is **Task Observation**. Instead of simply looking at the workplace in a general way, Task Observation focuses on how a specific job is actually done. This matters because hazards often appear in the details of movement, posture, sequence, timing, and behavior.

A task may look safe on paper, yet be performed in a risky way in reality because of speed, awkward position, worker habit, unsuitable equipment, or poor supervision. Observing the real task allows the business to identify hazards related to manual handling, repetitive motion, machine interaction, access problems, tool misuse, overreaching, unstable loading, and shortcut behavior. In small enterprises, where Workers often adapt tasks creatively to "get things done," Task Observation is especially valuable.

Review of past incidents, near misses, and complaints

A third technique is the **review of past incidents, near misses, and complaints**. Previous events often reveal hazard patterns that are still present in the workplace. If Workers have previously reported slips, cuts, strains, equipment faults, chemical splashes, fatigue, or minor collisions, those reports are clues that the hazards remain or may return.

Near misses are particularly useful because they show where serious harm nearly occurred, often without the cost of actual injury. Small businesses sometimes ignore near misses because no one was hurt. That is a mistake. Near misses are warning flares, and based on based on risk assessment methods defined by the organization in line with ISO 45001:2018 Clause 6.1.2" to align with risk-based thinking. should treat them as such.

Worker consultation and interviews

Worker consultation and interviews are also essential techniques. Workers are often the people closest to the actual job, the actual equipment, and the actual difficulties of the workday. They know where things jam, where loads are heavier than they should be, which tools are unreliable, where shortcuts happen, and which controls are impractical.

Consulting Workers through discussions, meetings, toolbox talks, informal reporting, or direct interviews often reveals hazards that management may not notice. For small businesses, this technique is powerful because communication lines are usually shorter and less formal. A business that listens well can learn a great deal quickly.

Review of equipment, materials, and process information

A further technique is the **review of equipment, materials, and process information**. Hazards are often linked to what the business uses. Machinery manuals, maintenance records, safety data sheets for chemicals, supplier instructions, inspection records, and manufacturer warnings all contain hazard information.

Reviewing these sources helps identify risks related to equipment failure, unsafe operating conditions, chemical exposure, maintenance needs, thermal hazards, electrical hazards, and incompatibility between materials or tools. In small businesses, such information is often available but not systematically reviewed. That creates a gap between what is known in documents and what is controlled in practice.

Change Analysis

Change Analysis is another important technique. Many hazards appear when something changes. This may include new machinery, new materials, changes in workflow, changes in staffing, increased production demand, new shifts, layout changes, new contractors, or relocation of tasks.

A small business should ask before any significant change: what new hazards could this introduce, and how will current risks be affected? This technique is particularly valuable because change often produces rushed decisions, and rushed decisions are fertile soil for risk.

Job hazard analysis

Small businesses can also use **job hazard analysis**, sometimes called task-based hazard breakdown. This technique divides a task into steps and examines each step for possible hazards. It is especially useful where tasks are repetitive, technical, involve equipment interaction, or include several stages with different risk levels.

For example, a maintenance task may involve isolation, access, tool use, removal, repair, testing, and restart. Each stage can have its own hazards. Breaking the task down prevents hazards from being hidden inside the overall activity.

Checklist-based based on based on risk assessment methods defined by the organization in line with ISO 45001:2018 Clause 6.1.2" to align with risk-based thinking.

Another useful technique is **checklist-based based on based on risk assessment methods defined by the organization in line with ISO 45001:2018 Clause 6.1.2" to align with risk-based thinking.**. Checklists help small businesses review common categories of hazards in a structured way. They can cover slips and trips, fire, electrical safety, machinery, chemicals, storage, PPE, ergonomics, manual handling, emergency arrangements, welfare facilities, and contractor control.

Checklists are helpful because they improve consistency and reduce the chance of forgetting obvious areas. However, they should not replace thinking. A checklist is a support tool, not a substitute for observation and judgment.

Technical Assessments

Where specialist hazards exist, small businesses may also need **Technical Assessments**. These may include noise measurements, air quality reviews, electrical inspections, structural checks, lifting equipment examinations, ergonomic reviews, or fire risk assessments performed by competent persons.

Not every small business needs advanced technical assessment for every activity, but where the hazard demands specialist knowledge, it is wiser to obtain proper advice than to improvise with confidence and regret.

Tools for Evaluating and Managing Risks in Small Enterprises

Once **hazards** have been identified, the next step is to evaluate the **risks** and decide how they will be managed. **Based on based on risk assessment methods defined by the organization in line with ISO 45001:2018 Clause 6.1.2" to align with risk-based thinking.** tells the business what can cause harm. **Risk assessment** tells the business how serious that hazard is in its specific conditions and what level of control is needed.

Risk Assessment Matrix

One of the most widely used tools for evaluating **risk** is the **risk assessment matrix**. This tool typically combines two factors: the likelihood of the event and the severity of the harm. By assigning values to these factors, the business can determine whether a **risk** is low, medium, high, or very high.

For small enterprises, the matrix is useful because it provides a simple and visual way to compare **risks** and prioritize action. It helps management decide where immediate attention is required and where existing controls may already be adequate. Its usefulness depends on consistent judgment and realistic scoring.

Hierarchy of Controls

Another important tool is the **hierarchy of controls**. This is not just a theory; it is a practical decision tool for selecting the most effective control measures. The hierarchy moves from strongest to weakest control:

1. **Elimination of the hazard**
2. **Substitution with something safer**
3. **Engineering controls**
4. **Administrative controls**
5. **Personal protective equipment (PPE)**

For example, if a chemical is hazardous, the strongest option may be to eliminate it or replace it with a safer one. If a machine creates a dangerous moving part, **engineering controls** such as machine guarding are stronger than relying only on worker caution. In small enterprises, the **hierarchy of controls** is extremely useful because it prevents overdependence on weak controls. Many businesses rush straight to PPE because it seems quick and visible. The hierarchy reminds the organization to ask whether stronger controls are possible first.

Risk Control or Action Plan

A very practical management tool is the **action plan** or **risk control plan**. Once **risks** are evaluated, the organization must decide what action will be taken, by whom, and by when. The **action plan** records these decisions clearly. It may include the hazard, existing controls, additional actions required, responsible person, due date, and status.

This turns assessment into management. Without an **action plan**, **risk assessment** can remain an intellectual exercise that changes nothing.

Job Hazard Analysis Forms

Job Hazard Analysis (JHA) forms are also useful as both **evaluation** and **management** tools. They allow the business to document the steps of a task, the **hazards** in each step, and the control measures required before the task begins.

This is particularly useful for high-risk, non-routine, maintenance, or contractor-related work. In small enterprises, such forms can help standardize control where work varies or where supervision is not constant.

Inspection Checklists

Another tool is the **inspection checklist** used for ongoing **risk** monitoring. Once controls are introduced, the business needs a way to verify that they remain in place and effective. **Inspection checklists** can cover housekeeping, machine guards, storage, signage, PPE use, electrical condition, emergency exits, fire equipment, spill control, and general workplace condition.

These tools help monitor whether **risk controls** are actually being maintained over time.

Incident Investigation Forms

The **incident investigation form** is also an important **risk management** tool. When something goes wrong, the business should not stop at recording the event. It should investigate the cause, identify what failed, and determine what corrective actions are needed.

Incident investigation turns events into learning. It strengthens future **risk control** by examining root causes rather than blaming only immediate behavior.

Training Matrices

Training matrices are valuable tools for **risk management** because many **risks** exist not only because of physical hazards but because of gaps in competence. A **training matrix** helps the business track who needs what training, what has been completed, and what remains overdue.

This helps management ensure that Workers are not assigned to tasks without the required competence.

Maintenance schedules and preventive maintenance logs

For some businesses, **maintenance schedules and preventive maintenance logs** are crucial tools for managing risk. Equipment-related hazards often increase when maintenance is delayed, irregular, or undocumented. A planned maintenance log helps keep machinery, vehicles, tools, and safety systems in safe condition.

This is especially important in small businesses where equipment may be used heavily and replaced less often.

Chemical registers and Safety Data Sheets

Where chemicals are used, **Safety Data Sheets and chemical registers** become important risk tools. These provide information on hazards, storage requirements, personal protection, emergency response, and incompatibilities.

Small enterprises using chemicals should not rely on label memory or enthusiastic guessing. The documentation should be available, understood, and used in control decisions.

Legal and compliance registers

A further practical tool is the **legal and compliance register**. This helps the business evaluate whether the controls it applies meet relevant legal obligations. Risk management is not only about what seems sensible; it must also align with what is required by law.

A compliance register supports this by linking hazards and controls to applicable duties.

Performance indicators

For ongoing review, the business may also use **performance indicators** such as number of inspections completed, percentage of overdue corrective actions, incident frequency, near-miss reports, training completion, or maintenance backlog. These indicators help management evaluate whether risk controls are functioning over time.

In small enterprises, simple indicators are often best because they are easier to maintain and more likely to be used.

Table: Techniques for Identifying Hazards and Tools for Evaluating and Managing Risks in Small Enterprises

Category	Technique / Tool	Purpose	Practical Use in Small Enterprises
Based on based on risk assessment methods defined by the organization in line with ISO 45001:2018 Clause 6.1.2" to align with risk-based thinking.	**Workplace Walk-Through Inspection**	Identify visible hazards in the work area	Useful for spotting poor housekeeping, blocked exits, machine issues, storage problems, and environmental hazards
Based on based on risk assessment methods defined by the organization in line with ISO 45001:2018 Clause 6.1.2" to align with risk-based thinking.	**Task Observation**	Identify hazards in the way work is actually performed	Helps reveal unsafe methods, awkward postures, shortcut behavior, and hidden task-related risks
Based on based on risk assessment methods defined by the organization in line with ISO 45001:2018 Clause 6.1.2" to align with risk-based thinking.	**Incident and near-miss review**	Identify recurring hazards and weak controls	Uses previous events as warning signs for current risk
Based on based on risk assessment methods defined by the organization in line with ISO 45001:2018 Clause 6.1.2" to align with risk-based thinking.	**Worker consultation and interviews**	Capture practical knowledge from Workers	Useful for understanding real work conditions and hidden hazards
Based on based on risk assessment methods defined by the organization in line with	**Review of manuals, SDS, records, and**	Identify hazards linked to equipment,	Supports understanding of technical and chemical hazards

ISO 45001:2018 Clause 6.1.2" to align with risk-based thinking.	**supplier information**	substances, and processes	
Based on based on risk assessment methods defined by the organization in line with ISO 45001:2018 Clause 6.1.2" to align with risk-based thinking.	**Change Analysis**	Identify hazards introduced by change	Useful before new equipment, new layouts, or new work methods are introduced
Based on based on risk assessment methods defined by the organization in line with ISO 45001:2018 Clause 6.1.2" to align with risk-based thinking.	**Job hazard analysis**	Break tasks into steps and identify step-specific hazards	Useful for technical, repetitive, or high-risk tasks
Based on based on risk assessment methods defined by the organization in line with ISO 45001:2018 Clause 6.1.2" to align with risk-based thinking.	**Checklist-based review**	Ensure hazard categories are reviewed consistently	Helps small businesses avoid missing common risk areas
Risk Evaluation / Management	**Risk assessment matrix**	Evaluate severity and likelihood of risks	Helps prioritize actions and compare risk levels
Risk Evaluation / Management	**Hierarchy of controls**	Select the most effective control measures	Prevents overreliance on weak controls such as PPE alone
Risk Evaluation / Management	**Risk control / action plan**	Assign and track control actions	Ensures identified risks lead to real action
Risk Evaluation / Management	**Job Hazard Analysis form**	Document task hazards and required controls	Useful before non-routine or higher-risk activities
Risk Evaluation / Management	**Inspection checklist**	Monitor the condition of controls over time	Helps confirm controls remain in place and effective

Risk Evaluation / Management	**Incident investigation form**	Analyze causes and determine corrective actions	Supports learning and prevention after events
Risk Evaluation / Management	**Training matrix**	Manage competence-related risks	Ensures Workers receive training relevant to their tasks
Risk Evaluation / Management	**Maintenance log / preventive maintenance plan**	Control equipment-related risk	Helps reduce failure, malfunction, and unsafe equipment conditions
Risk Evaluation / Management	**Chemical register / Safety Data Sheets**	Manage substance-related risks	Supports safer handling, storage, and emergency response
Risk Evaluation / Management	**Legal compliance register**	Link risk controls with legal duties	Helps confirm regulatory requirements are addressed
Risk Evaluation / Management	**Performance indicators**	Monitor effectiveness of risk management over time	Helps management review trends and system performance

Conclusion

For small businesses, based on based on risk assessment methods defined by the organization in line with ISO 45001:2018 Clause 6.1.2" to align with risk-based thinking. and risk management must be practical, structured, and close to daily operations. Based on based on risk assessment methods defined by the organization in line with ISO 45001:2018 Clause 6.1.2" to align with risk-based thinking. techniques help the organization see where harm may arise. Risk evaluation and management tools help the organization judge seriousness, choose suitable controls, assign actions, and monitor effectiveness.

Together, these techniques and tools give the small enterprise something extremely valuable: visibility, control, and direction. Without them, the business is left relying on memory, habit, and luck. Luck is charming in card games. It is much less impressive in Occupational Health and Safety management.

Author's Personal Note

Risk assessment and based on based on risk assessment methods defined by the organization in line with ISO 45001:2018 Clause 6.1.2" to align with risk-based thinking. are often spoken of as technical requirements, yet their true value is deeply practical. They represent the point at which a business stops relying on routine, assumption, or experience alone, and begins to examine work with greater discipline and responsibility. For small enterprises, this shift is particularly important. Limited resources and closely shared duties can make it easy for hazards to become part of normal working life, unnoticed until an incident forces attention upon them.

What this chapter seeks to emphasize is that effective risk assessment does not depend on complicated systems or excessive documentation. Its strength lies in careful observation, honest evaluation, and timely action. When a small business develops the habit of identifying hazards early and assessing risks thoughtfully, it strengthens not only compliance, but also the stability and resilience of the organization itself.

Across all sectors, the principle remains the same: harm is rarely prevented by accident. It is prevented by awareness, structure, and consistent management attention.

The intention behind this chapter is therefore not to burden businesses with additional formality, but to encourage a more practical and informed way of managing health and safety. Even simple improvements in based on based on risk assessment methods defined by the organization in line with ISO 45001:2018 Clause 6.1.2" to align with risk-based thinking. and risk evaluation can create meaningful change when applied consistently. In that sense, the purpose of this chapter is not only to explain process, but to support a more thoughtful safety culture within the day-to-day realities of small business operations.

Kind regards,
Dr. Mohamed-Ali Ibrahim
Vienna, Austria

CHAPTER 8

Legal and Compliance Requirements for Small Enterprises

Legal Obligations for OH&S in Small Enterprises

Legal obligations in Occupational Health and Safety are not optional, and they do not become optional merely because a business is small, busy, or running on thin margins. A small enterprise may have fewer Workers, fewer departments, and fewer layers of management, but it still has legal responsibilities to protect Workers and other persons who may be affected by its activities. In practical terms, legal compliance means the business must understand what health and safety laws apply to its operations, put suitable controls in place, monitor whether those controls are working, and correct any failures before they lead to harm or enforcement action.

For small enterprises, legal obligations can sometimes feel scattered, technical, or difficult to interpret. Rules may come from labor law, workplace safety law, fire safety regulations, electrical safety requirements, machinery provisions, first aid duties, chemical handling rules, welfare requirements, working time obligations, contractor controls, transport rules, and sector-specific regulations. Because of this, some small businesses fall into a dangerous habit of assuming that "basic common sense" is enough. It is not. Common sense helps, but legal compliance requires a more structured approach.

Why legal compliance matters in small enterprises

Legal compliance matters because the law is designed to prevent harm before it occurs. It requires employers and business operators to think ahead, manage foreseeable risks, and create safe working conditions. When a small enterprise fails to meet legal requirements, the consequences may include injury, illness, prosecution, fines, insurance complications, enforcement notices, reputational damage, loss of contracts, and disruption to normal business operations.

For a small business, these consequences may be severe because there is often less financial and operational capacity to absorb them. One enforcement action or one serious incident can create pressure across the whole organization. That is why legal compliance should not be treated as an afterthought. It should be understood as a core part of business responsibility and sustainability.

Legal compliance also matters because it supports the credibility of the OH&S Management System. A business cannot claim to manage safety effectively if it does not know its own legal duties. ISO 45001:2018 expects the organization to determine applicable legal requirements and other requirements, integrate them into the OH&S Management System, and evaluate compliance over time. Legal control is therefore not separate from the management system; it is part of its backbone.

The general duty of care

Most Occupational Health and Safety laws are built around a broad principle often called the **general duty of care**. While the exact wording differs from one country to another, the basic idea is similar: the employer or responsible business operator must provide, as far as reasonably practicable, a safe and healthy work environment.

This general duty usually includes responsibility for:

- safe workplaces and access routes
- safe equipment, tools, and machinery
- safe systems of work
- information, instruction, training, and supervision
- suitable welfare arrangements
- control of hazardous substances and conditions
- emergency preparedness
- protection of Workers and others affected by the work

For small enterprises, this means that safety cannot be left entirely to worker judgment or personal caution. The business itself has a legal duty to create conditions in which safe work is possible and expected.

Common legal obligation areas for small enterprises

Although exact legal duties depend on national and sector-specific regulations, small enterprises commonly face OH&S obligations in several recurring areas.

Legal and Compliance Requirements Framework for Small Enterprises

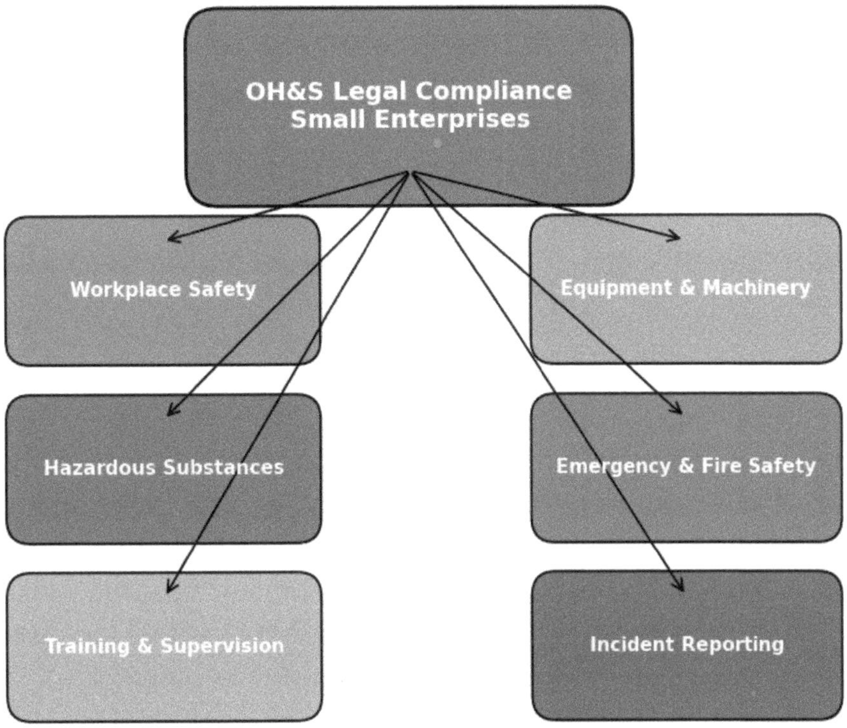

Workplace safety and conditions

The business must provide a workplace that is safe and without unnecessary risk. This often includes maintaining safe floors, walkways, lighting, ventilation, temperature control, access and egress, housekeeping, sanitation, and emergency exits. Poor physical conditions may seem ordinary when they develop gradually, but the law does not treat blocked exits, slippery floors, exposed wiring, or unsafe storage as charming personality traits of a busy business.

Machinery, tools, and equipment

Where machinery or equipment is used, the organization usually has legal duties related to guarding, inspection, maintenance, safe operation, and worker instruction. Unsafe tools and poorly maintained machines are common sources of injury in small enterprises. Legal compliance requires more than buying equipment and hoping everyone behaves sensibly around it.

Manual handling and ergonomic risks

Many small businesses rely heavily on lifting, carrying, pushing, pulling, repetitive work, awkward postures, and prolonged standing or sitting. These activities can cause strain injuries, chronic pain, and lost working capacity. Legal duties often require the business to assess these risks and reduce them where reasonably practicable through redesign, mechanical aids, training, job variation, or other controls.

Hazardous substances

If the business uses chemicals, cleaning agents, fuels, paints, solvents, or other hazardous substances, legal obligations usually include identification of hazards, safe storage, labeling, availability of safety data information, exposure control, training, spill response, and proper disposal. Chemical hazards must be identified, labeled, and managed according to regulatory and ISO 45001 requirements."

Fire and emergency arrangements

Most businesses have legal duties related to emergency preparedness, including fire detection or alarms where required, emergency exits, fire-fighting equipment, evacuation arrangements, and worker awareness. In some sectors, the law may require written emergency plans, drills, and formal fire risk assessments. A small enterprise must determine what level of emergency preparation is legally required for its type of work and premises.

First aid and welfare

The business may be legally required to provide first aid arrangements, trained first aid personnel where appropriate, and welfare facilities such as drinking water, washrooms, rest arrangements, and basic sanitation. These duties support health as well as safety and are often overlooked when management focuses only on machines and visible hazards.

Training, instruction, and supervision

One of the strongest recurring legal duties is to provide Workers with appropriate information, instruction, training, and supervision. The level of training should match the risks and the nature of the work. A worker cannot lawfully be expected to perform hazardous or unfamiliar work safely without being properly informed and prepared.

Contractor and visitor safety

Where contractors, delivery personnel, clients, or other visitors enter the workplace, the business may have duties to control shared risks, provide relevant safety information, restrict access where needed, and coordinate work activities. A small enterprise does not lose responsibility simply because another company or outside person is present on site.

Incident reporting and record keeping

Many legal systems require certain incidents, injuries, dangerous occurrences, or occupational diseases to be reported to authorities. Businesses may also be required to keep records of injuries, inspections, maintenance, training, exposure, or compliance checks. These records help demonstrate that safety duties are being managed systematically.

How small enterprises should determine their legal obligations

One of the biggest compliance problems in small businesses is not deliberate neglect but lack of a structured method for identifying what applies. Legal obligations should not be discovered only after an inspector, auditor, insurer, or lawyer appears at the door with an unpleasant expression.

A small enterprise should establish a process to determine applicable legal and other requirements. This process may include reviewing national legislation, local authority guidance, industry regulations, insurance requirements, landlord conditions, supplier instructions, professional association guidance, and client-specific safety requirements where relevant. The business should identify which laws apply to its activities, premises, equipment, workforce, substances, vehicles, and work methods.

This process should be kept practical. It does not require turning the business owner into a part-time legal scholar living under fluorescent light. It does require a clear and repeatable method for identifying, reviewing, and updating applicable requirements.

Integrating legal requirements into the OH&S Management System

Knowing the law is not enough. The business must integrate legal obligations into the OH&S Management System so that compliance is reflected in actual operations.

This means legal requirements should influence:

- risk assessments
- operational procedures
- maintenance schedules
- training plans
- contractor controls
- emergency arrangements
- inspection routines
- incident reporting processes
- management review
- corrective action systems

For example, if legal rules require machine guarding, then guarding must appear not only in a legal register but also in purchase standards, inspections, maintenance controls, and worker training. If chemical rules require labels and safety data sheets, then those requirements must appear in procurement, storage, handling procedures, and training. If first aid coverage is legally required, that duty must be reflected in staffing, awareness, and emergency arrangements.

This is how compliance becomes operational rather than theoretical.

Evaluating compliance over time

A small enterprise should not assume that once legal obligations are identified, compliance will maintain itself politely in the background. Conditions change. Equipment changes. Premises change. Staff change. Laws and guidance also change. The business must therefore evaluate compliance periodically.

Compliance evaluation may include:

- reviewing whether legal duties have been identified correctly
- checking whether required controls are actually in place
- verifying records, inspections, and maintenance activities
- reviewing incidents for signs of legal failure
- confirming that required training has been completed
- examining whether emergency arrangements remain adequate
- updating obligations when laws or work conditions change

Under ISO 45001:2018, compliance evaluation is an expected part of the management system. It is not enough to have good intentions and a dusty binder of regulations. The organization must maintain knowledge and understanding of its compliance status.

Responsibilities for legal compliance

In a small enterprise, legal compliance responsibilities must be clearly assigned. Ultimate accountability normally rests with top management or the business owner. However, day-to-day compliance tasks may be shared among managers, supervisors, safety coordinators, maintenance personnel, HR support, or external advisers depending on the size and complexity of the business.

Typical responsibilities may include:

- identifying applicable laws and requirements
- updating the legal register
- ensuring required inspections and maintenance are completed
- verifying training obligations are met
- reviewing contractor compliance
- supporting incident reporting
- evaluating compliance Observations and recommendations must be documented; implementation is per organization discretion.
- initiating corrective actions when gaps are found

The key point is clarity. If everyone vaguely assumes someone else is "probably handling the legal stuff," then eventually no one is handling it properly.

Common compliance weaknesses in small enterprises

Small businesses often face recurring weaknesses in OH&S legal compliance. These usually do not arise from evil intent. They arise from pressure, limited time, limited expertise, and informal working habits. Still, the law tends not to be moved by heartfelt explanations about being very busy.

Common weaknesses include:

- not knowing which legal duties apply
- relying on outdated legal information
- failing to review changes in law or operations
- weak maintenance and inspection records
- poor contractor control
- incomplete training documentation
- unsafe chemical storage or labeling
- blocked emergency exits or weak fire arrangements
- informal reporting of incidents with no records
- assuming that long-standing practice equals compliance

Recognizing these patterns helps the enterprise take corrective action before enforcement pressure or incidents force the matter.

Benefits of meeting legal obligations effectively

Effective legal compliance gives the business more than protection from fines or enforcement. It strengthens the entire management system. It improves worker confidence, supports operational consistency, reduces incident likelihood, improves audit performance, and strengthens the organization's credibility with clients, regulators, and insurers.

It also supports:

- better risk control
- clearer responsibilities
- stronger training discipline
- more reliable records
- better preparedness for inspections and audits
- greater confidence in business continuity

For a small enterprise, this is valuable. Compliance done properly is not dead weight. It is a stability mechanism.

Key practical points for small enterprises

A small enterprise should remember several basic truths about OH&S legal obligations:

- legal compliance must be structured, not assumed
- the size of the business does not remove duty of care
- legal duties must be identified, applied, and reviewed
- compliance should be integrated into daily operations
- records and evidence matter
- Workers need information, training, and supervision
- changes in business activity should trigger legal review
- compliance is an ongoing process, not a one-time event

These points may seem straightforward, but they are where many businesses either build control or wander into avoidable trouble.

Conclusion

Legal obligations for OH&S in small enterprises form a core part of responsible business management. They require the organization to provide safe working conditions, control hazards, train Workers, manage emergencies, maintain suitable records, and monitor whether legal requirements are actually being met. For small businesses, these duties are especially important because the effect of one failure can spread quickly across the whole operation.

A strong OH&S Management System under ISO 45001:2018 does not treat legal compliance as a side topic. It treats it as an active, structured, and continually reviewed part of the system. When small enterprises understand their legal obligations and build them into everyday operations, they strengthen both worker protection and business stability.

The law, rather inconveniently for those who enjoy improvisation, expects evidence rather than optimism.

Staying Updated with OH&S Regulations for Small Businesses

Small enterprises operate in a dynamic environment where health and safety regulations can change frequently. Staying up-to-date with OH&S regulations is critical to ensure that legal obligations are met, risks are controlled, and the business remains compliant under ISO 45001:2018. Small businesses often face challenges in this area because they may not have dedicated safety officers or legal advisors, making it easy to miss changes in legislation, guidance, or industry standards.

Staying updated is not just about compliance—it also protects Workers, avoids penalties, strengthens credibility with clients and regulators, and ensures smooth operational continuity.

Key Steps to Staying Updated

- **Subscribe to Regulatory Updates**: Register for notifications from government safety authorities, labor departments, and industry associations.

- **Use Professional Networks**: Engage with trade groups, professional associations, and local business networks to receive timely guidance.
- **Schedule Periodic Reviews**: Assign a regular schedule (quarterly or semi-annually) to review legal obligations, safety bulletins, and regulatory changes.
- **Document Changes**: Keep a log of updates, including date, source, impact, and required actions for the business.
- **Train Staff on Changes**: Ensure that Workers and supervisors are informed about relevant regulatory changes that affect their work.

Creating and Using a Compliance Matrix in Small Enterprises

A **compliance matrix** is a practical tool that allows small businesses to map OH&S regulations against the organization's processes, responsibilities, and control measures. It provides a visual summary of which requirements apply, who is responsible for each, and how compliance is achieved. This tool simplifies the management of legal obligations and helps small businesses demonstrate due diligence during audits.

Benefits of a Compliance Matrix

- Provides a clear overview of legal obligations.
- Identifies responsibilities and accountability for compliance.
- Ensures that all regulations are linked to actual operational controls.
- Highlights gaps in current compliance or documentation.
- Supports audit preparation and ISO 45001 certification readiness.

Steps to Create a Compliance Matrix

1. **List Applicable Regulations**: Identify all relevant OH&S laws, regulations, codes, and guidance affecting the business.
2. **Map to Processes**: Determine which parts of the business operations each requirement affects.
3. **Assign Responsibilities**: Specify who in the organization is responsible for compliance.

4. **Document Controls**: Identify existing measures and procedures that address each regulation.
5. **Monitor Status**: Indicate whether each requirement is fully compliant, partially compliant, or requires action.
6. **Review Regularly**: Update the matrix whenever regulations change or new processes are introduced.

Example Compliance Matrix for a Small Enterprise

Regulation / Requirement	Business Process Affected	Responsible Person	Current Control / Procedure	Compliance Status	Next Review Date
Fire Safety Regulations	Warehouse, Office	Operations Manager	Fire exits marked, fire drills scheduled quarterly	Compliant	
Machinery Safety Standards	Production / Workshop	Maintenance Supervisor	Machine guards installed, inspection logs maintained	Partial	
Chemical Handling Regulations	Cleaning, Workshop	Safety Officer	SDS available, PPE provided, staff trained	Compliant	
Manual Handling Requirements	Delivery & Stock Handling	Warehouse Supervisor	Training, lifting aids, regular ergonomic review	Compliant	
First Aid Provision	All work areas	HR Manager / First Aid Officer	First aid kits checked monthly, trained personnel available	Compliant	

Practical Tips for Small Enterprises

- Keep the matrix simple and manageable; too much complexity reduces usability.

- Use color-coding to quickly highlight compliance status: green = compliant, yellow = partially compliant, red = needs action.
- Link each matrix item to documentation or evidence in the OH&S Management System.
- Assign someone responsible for maintaining and updating the matrix regularly.
- Integrate the matrix into audits and management reviews to demonstrate active compliance monitoring.

Conclusion

For small businesses, staying current with OH&S regulations and using a compliance matrix are practical, low-cost ways to maintain legal compliance and operational control. The compliance matrix provides a single reference point linking laws, responsibilities, and operational measures. Together with regular monitoring and training, these practices help small enterprises manage legal obligations effectively, protect Workers, and prepare for ISO 45001 audits.

Author's Personal Note

Understanding and managing legal and compliance requirements can feel overwhelming for small enterprises, especially when regulations change frequently and resources are limited. However, staying informed and organized is not just a legal necessity—it is a practical strategy for protecting Workers, reducing risk, and maintaining business stability.

The tools and approaches described in this chapter, including **staying updated with regulations** and **creating a compliance matrix**, are intended to make compliance manageable, transparent, and integrated into daily operations. A compliance matrix is not merely a record; it is a living tool that helps a small business track obligations, assign responsibilities, monitor effectiveness, and quickly respond to changes.

Small enterprises may not have large safety teams or legal departments, but with structure, simple tools, and regular review, compliance can be maintained efficiently. This chapter encourages business owners and managers to treat legal obligations as part of operational management rather than as a separate, overwhelming burden.

Remember, legal compliance in OH&S is not just about avoiding penalties—it is about creating a safe and healthy work environment, demonstrating responsibility, and fostering trust with Workers, clients, and regulators. Even small, practical steps taken consistently will have a significant impact on workplace safety and business resilience.

Kind regards,
Dr. Mohamed-Ali Ibrahim
Vienna, Austria

CHAPTER 9

Training and Competency in OH&S for Small Enterprises

Developing an OH&S Training Program for Small Businesses

Developing an effective **Occupational Health and Safety (OH&S) training program** is one of the most critical steps in ensuring that small enterprise Workers can work safely and competently. While policies and procedures define what must be done, training ensures Workers know **how to apply these requirements in real operations**. A well-structured training program reduces incidents, strengthens the safety culture, and supports ISO 45001:2018 compliance.

In small businesses, Workers often perform multiple roles, supervision is limited, and work processes can change frequently. Without targeted OH&S training, even a simple task can present significant risks if Workers are unaware of hazards or proper controls. Developing a training program involves identifying needs, defining objectives, selecting appropriate methods, scheduling, assigning responsibility, and ensuring that the program is measurable and effective.

Key Steps to Develop an OH&S Training Program

1. **Identify Training Needs**
   - Review job roles, tasks, and operational hazards.
   - Include both routine tasks (daily work) and non-routine tasks (maintenance, emergency response).
   - Consider staff turnover, temporary staff, and contractors.
2. **Define Learning Objectives**
   - Clearly specify the knowledge and skills Workers must acquire.
   - Objectives should be measurable and directly linked to operational safety requirements.
   - Example: "Workers must be able to safely operate forklift machinery without risk of injury."
3. **Select Training Methods**

- Choose methods appropriate for the workforce and type of hazard:
  - Classroom sessions
  - Toolbox talks
  - On-the-job coaching
  - Demonstrations
  - Practical drills
  - E-learning modules for procedural or regulatory knowledge

4. **Schedule Training**
   - Incorporate induction for new hires.
   - Schedule refresher courses at regular intervals.
   - Provide additional training whenever new equipment, procedures, or hazards are introduced.
5. **Assign Responsibility**
   - Designate qualified trainers, supervisors, or safety officers to deliver and track training.
   - In very small businesses, the owner may take responsibility but must ensure accountability and documentation.
6. **Document Training Activities**
   - Maintain records for each employee, including topics covered, date, trainer, and assessment results.
   - Proper documentation supports audits, compliance verification, and continual improvement.

Tips for Developing an Effective Program in Small Enterprises

- **Focus on practical application:** Training should emphasize real workplace scenarios rather than just theory.
- **Use multiple methods:** Combine classroom, on-the-job, and hands-on training for better engagement and retention.
- **Prioritize high-risk tasks:** Ensure tasks with higher hazards receive more detailed instruction and supervision.
- **Integrate with operational schedules:** Training should not disrupt daily operations but align with workflow.
- **Monitor and adjust:** Continuously evaluate training effectiveness and adjust content or methods as needed.

Benefits of a Structured OH&S Training Program

- Reduces workplace incidents and injuries.
- Increases employee confidence and competence.
- Ensures compliance with ISO 45001:2018 and local OH&S regulations.
- Strengthens the organization's safety culture.
- Provides measurable records for audits and management reviews.

Example Training Needs Matrix for Small Businesses

Employee Role	Training Required	Frequency	Responsible Trainer/Supervisor	Evaluation Method
Production Operator	Machine operation, PPE, hazard recognition	Induction + Annual refresher	Production Supervisor	Observation + Quiz
Warehouse Staff	Manual handling, forklift, storage safety	Induction + Semi-annual refresher	Warehouse Manager	Observation + Practical test
Cleaner / Maintenance Staff	Chemical handling, PPE, emergency response	Induction + Annual refresher	Safety Officer	Observation + Checklist
Office Staff	Ergonomics, emergency evacuation	Induction + Annual refresher	HR Manager	Observation + Feedback
Supervisors	Based on based on risk assessment methods defined by the organization in line with ISO 45001:2018 Clause 6.1.2" to align with risk-based thinking., risk assessment, leadership	Induction + Annual refresher	Safety Officer	Observation + Audit review
Contractors	Site safety induction, emergency procedures	Before work + Project refresh	Site Supervisor	Observation + Checklist

Conclusion

A well-developed OH&S training program ensures that Workers in small enterprises are competent, aware of hazards, and able to follow safe work practices. By identifying training needs, defining clear objectives, selecting appropriate methods, and maintaining proper documentation, small businesses can build a workforce capable of managing risk effectively, supporting compliance, and fostering a safety-focused workplace culture.

Training is not just a compliance requirement—it is a practical tool that protects Workers, strengthens operations, and lays the foundation for a sustainable OH&S Management System.

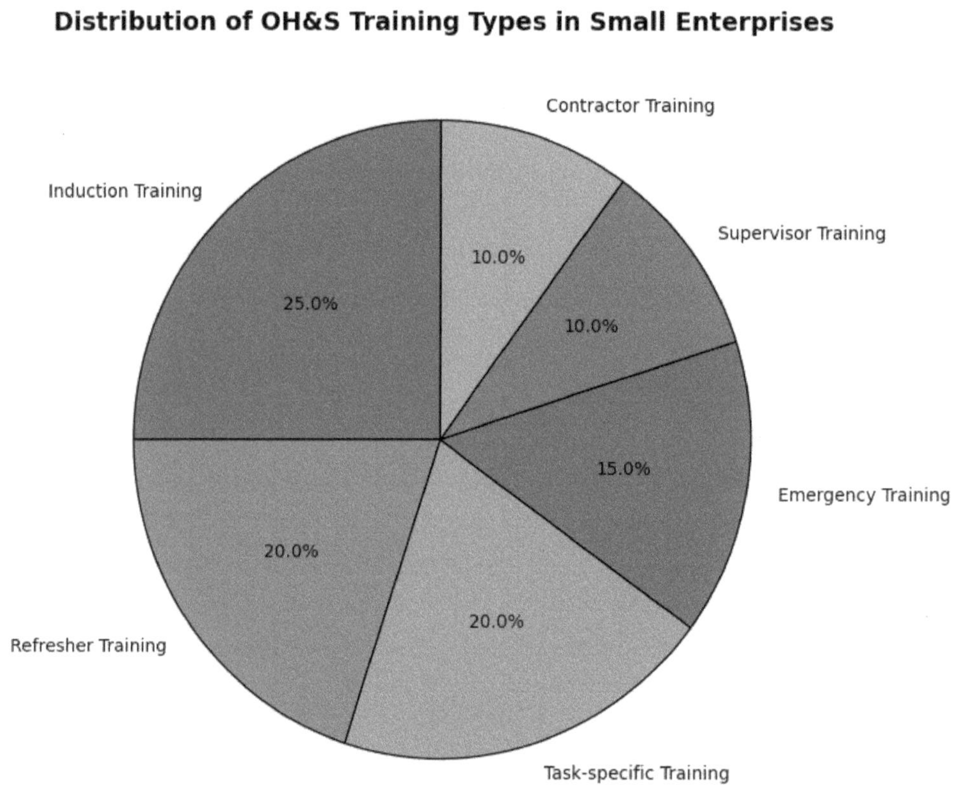

Ensuring Ongoing Competency and Skill Development in Small Enterprises

Ensuring ongoing competency is a crucial component of an effective **Occupational Health and Safety (OH&S) Management System** in small enterprises. Unlike one-off training sessions, ongoing competency ensures that Workers maintain the knowledge, skills, and awareness necessary to perform tasks safely, adapt to operational changes, and respond effectively to emergencies.

In small businesses, Workers often perform multiple roles and tasks, meaning that their exposure to hazards can vary widely. Without continuous skill development, gaps in knowledge or procedure can develop quickly, increasing the risk of incidents and non-compliance with ISO 45001:2018 requirements. Ongoing competency initiatives ensure that staff remain confident, capable, and prepared for the risks they may encounter in the workplace.

Key Strategies for Maintaining Ongoing Competency

- **Regular Refresher Training**
    - Conduct periodic sessions to reinforce knowledge and update Workers on new procedures, equipment, or regulations.
- **Task-Specific Training**
    - Provide focused training whenever new tasks, equipment, or hazards are introduced.
- **Mentoring and Supervision**
    - Use experienced Workers to coach and monitor newer staff, reinforcing correct safety practices.
- **Skill Verification**
    - Observe Workers performing critical tasks to confirm they apply procedures and controls correctly.
- **Performance Integration**
    - Incorporate competency checks into routine performance reviews to identify gaps and provide corrective training.
- **Continuous Learning Culture**

- Encourage Workers to provide feedback, ask questions, and suggest improvements to promote engagement and knowledge sharing.

Ongoing competency is not only about compliance—it is about creating a workforce that is adaptable, confident, and proactive in identifying and controlling hazards.

Assessing Training Effectiveness in Small Businesses

Simply providing training does not guarantee that Workers have understood, retained, or can apply the knowledge and skills. Assessing training effectiveness ensures that the time, effort, and resources invested in training deliver real improvements in workplace safety and operational performance.

For small enterprises, assessment helps identify gaps, adjust future training, and provide evidence of competency for audits and ISO 45001 certification.

Methods for Assessing Training Effectiveness

- **Observation of Work Practices**
  - Supervisors or safety officers monitor Workers performing tasks to ensure correct procedures are being followed.
- **Knowledge Testing**
  - Written quizzes, verbal questioning, or practical assessments can verify understanding of critical procedures, hazards, and controls.
- **Feedback Collection**
  - Gather input from Workers on the clarity, relevance, and applicability of training.
- **Incident and Near-Miss Analysis**
  - Monitor whether training has reduced incidents, near misses, or unsafe practices.
- **Audits and Inspections**
  - Include training effectiveness as part of internal OH&S audits and workplace inspections.

- **Competency Tracking**
  - Maintain a record of **Workers'** training, evaluations, and observed skill performance to identify trends or recurring gaps.

Benefits of Effective Training Assessment

- Confirms that Workers are competent and able to apply safe work procedures.
- Identifies areas for improvement in the training program.
- Supports continuous improvement in safety culture and OH&S performance.
- Provides documented evidence for ISO 45001 audits.
- Reduces risk of incidents caused by human error or lack of knowledge.

Table: Training and Competency Assessment Methods

Method	Purpose	Frequency / Notes
Observation of work	Verify safe procedures are followed	Ongoing during daily operations
Written/Practical Tests	Assess knowledge retention	After training and periodically for refreshers
Feedback Surveys	Identify gaps in understanding	After each session or quarterly
Incident / Near-Miss Review	Evaluate practical application of training	Monthly or after events
Internal Audits / Inspections	Ensure training translates to compliance	Semi-annual or annual
Competency Records / Tracking	Monitor skill development over time	Updated continuously

Conclusion

Ongoing competency and assessment of training effectiveness are critical to small enterprises that want to ensure their OH&S Management System works in practice. Continuous skill development, coupled with rigorous evaluation, ensures Workers can perform tasks safely, respond to hazards effectively, and maintain compliance with ISO 45001:2018. By investing in these processes, small businesses create a capable, adaptable, and safety-conscious workforce, reducing risks and supporting operational resilience.

Author's Personal Note

Training and competency development are not just formalities—they are the practical lifeblood of a safe and compliant workplace. In small enterprises, Workers often perform multiple roles, work under time pressure, and face changing tasks daily. Without ongoing competency, even the best OH&S policies and procedures may fail in practice.

This chapter emphasizes that effective OH&S training is continuous, targeted, and measurable. It is not a one-time event but a structured process that ensures Workers maintain the knowledge, skills, and awareness needed to work safely. Assessing training effectiveness is equally important, as it confirms that learning translates into real, safe behaviors and identifies areas for improvement.

By committing to continuous training, skill development, and regular assessment, small enterprises can strengthen their safety culture, reduce incidents, and maintain compliance with ISO 45001:2018. Ultimately, a well-trained workforce is not only safer—it is more confident, efficient, and resilient.

Kind regards,
Dr. Mohamed-Ali Ibrahim
Vienna, Austria

CHAPTER 10

Employee Involvement and Engagement in OH&S for Small Enterprises

Employee involvement and engagement are fundamental pillars of a strong **Occupational Health and Safety (OH&S) Management System**, particularly in small enterprises. Policies, procedures, and risk controls are only effective when the people who carry out the work understand, commit to, and participate actively in safety practices. Without employee engagement, even the best-designed OH&S Management System can become a set of documents rather than a living, functioning system.

In small businesses, Workers often work closely with supervisors and owners, and communication lines are short. This creates a unique opportunity: employee participation can directly shape operational safety, hazard reporting, and procedural improvement. Engaged Workers are more likely to recognize hazards, suggest practical controls, follow safe work practices, and contribute to continuous improvement.

Building a Safety Culture in Small Enterprises

A **safety culture** is the collection of shared values, beliefs, and behaviors that determine how health and safety is managed and prioritized within the organization. In small enterprises, building a strong safety culture is both essential and achievable because Workers often have direct interactions with leadership and decision-makers.

Key Components of a Safety Culture

1. **Visible Leadership Commitment**
    - Owners and supervisors demonstrate that safety is a priority in all operational decisions.
    - Management visibly participates in safety initiatives, including inspections, training, and risk discussions.
2. **Open Communication**

   - Workers are encouraged to report hazards, near misses, and unsafe practices without fear of reprisal.
   - Clear communication channels are established for sharing safety information and updates.
3. **Worker Participation**
   - Workers are actively involved in risk assessments, based on based on risk assessment methods defined by the organization in line with ISO 45001:2018 Clause 6.1.2" to align with risk-based thinking., and problem-solving.
   - Safety committees or small working groups can facilitate input on safety improvements.
4. **Recognition and Reinforcement**
   - Safe behavior is recognized and rewarded, creating positive reinforcement.
   - Incidents or near misses are used as learning opportunities rather than solely for blame.
5. **Training and Competency**
   - Workers are regularly trained and kept aware of hazards, safe practices, and procedures.
   - Competency is reinforced through observation, coaching, and mentoring.
6. **Continuous Improvement**
   - Feedback from Workers informs updates to procedures, hazard controls, and training.
   - Safety practices are regularly reviewed, audited, and refined.

Practical Strategies for Employee Engagement in OH&S

- **Safety Meetings and Toolbox Talks**
  - Short, focused discussions on hazards, controls, and safe work practices.
  - Encourages participation and reinforces awareness.
- **Suggestion Systems**
  - Workers can propose improvements to work practices, equipment use, or hazard controls.
- **Involvement in Risk Assessments**

- Workers can help identify hazards in their own tasks, ensuring practical controls are applied.
- **Peer-to-Peer Coaching and Mentoring**
    - Experienced Workers guide newer staff on safe practices.
    - Encourages knowledge transfer and reinforces a culture of safety.
- **Regular Feedback Loops**
    - Management provides feedback on safety concerns raised by Workers.
    - Ensures that employee input leads to tangible improvements.
- **Emergency Drills Participation**
    - Workers practice response to emergencies, enhancing preparedness and confidence.

Employee Engagement and Safety Culture in Small Enterprises

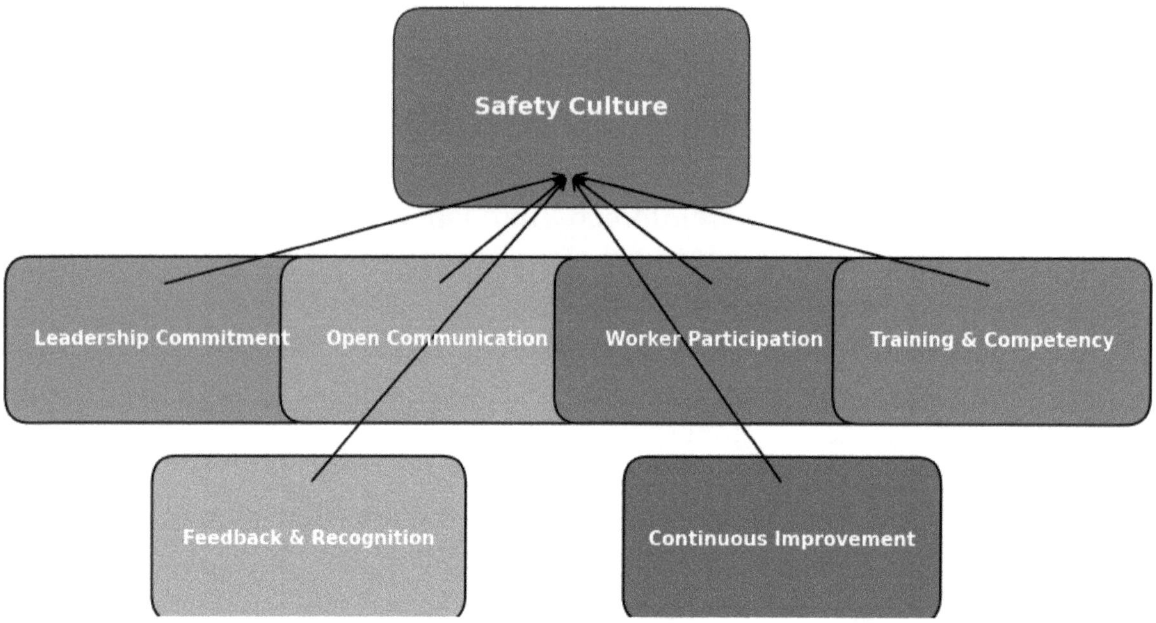

Table: Examples of Employee Engagement Activities in Small Enterprises

Engagement Activity	Purpose / Outcome	Frequency	Responsible Person
Toolbox Talks	Raise awareness of specific hazards and safe practices	Weekly or bi-weekly	Supervisor / Safety Officer

Safety Committee Meetings	Discuss hazards, incidents, and improvements	Monthly	Manager + Employee reps
Suggestion Box / Digital Form	Encourage employee input on safety improvements	Ongoing	Safety Officer / Manager
Risk Assessment Participation	Include Workers in based on based on risk assessment methods defined by the organization in line with ISO 45001:2018 Clause 6.1.2" to align with risk-based thinking. and risk evaluation	Annually or before changes	Supervisor / Worker Teams
Peer Mentoring	Transfer knowledge and safe practices between Workers	Ongoing	Experienced Worker / Supervisor
Emergency Drill Participation	Test and improve emergency response	Quarterly or semi-annual	Manager / Safety Officer
Incident Review and Feedback	Discuss causes and prevention measures with staff	After incidents or near misses	Supervisor / Safety Officer

Benefits of Employee Involvement in OH&S

- Increases hazard awareness across all levels of the business.
- Improves adherence to procedures and controls.
- Encourages proactive identification of hazards before incidents occur.
- Strengthens overall safety culture and employee morale.
- Enhances audit readiness and ISO 45001 compliance.
- Reduces workplace incidents and associated costs.

Conclusion

In small enterprises, building a safety culture and engaging Workers in OH&S activities is critical for effective risk management. Employee involvement ensures that hazards are identified, controls are applied correctly, and safe practices become part of daily work. Active participation transforms safety from a set of rules into a shared value, creating a workplace where safety is understood, practiced, and continuously improved.

When leadership commits, communication flows openly, and Workers are empowered to participate, small enterprises can achieve a sustainable, resilient OH&S culture that supports both compliance and operational excellence.

Encouraging Employee Participation in OH&S Activities in Small Businesses

Employee participation is a cornerstone of an effective **Occupational Health and Safety (OH&S) Management System**, especially in small enterprises where staff often work closely with supervisors and management. Participation ensures that Workers are not merely following rules, but are actively engaged in identifying hazards, suggesting improvements, and contributing to a safer workplace.

Engaged Workers help create a culture where safety becomes a shared responsibility rather than a top-down directive. This involvement improves hazard recognition, increases adherence to procedures, reduces incidents, and supports ISO 45001:2018 compliance. Small businesses benefit because participatory practices often reveal practical insights into risks that management alone may overlook.

Strategies to Encourage Participation

- **Involve Workers in risk assessments**
  - Workers contribute practical knowledge about hazards and operational risks in their daily tasks.
- **Create safety committees or working groups**
  - Small groups can discuss issues, propose solutions, and review incidents or near misses.
- **Implement suggestion systems**
  - Workers can submit ideas for improving safety, either via a physical suggestion box or digital platform.
- **Conduct regular safety meetings or toolbox talks**
  - Encourage open discussions about hazards, controls, and safety improvements.
- **Recognize and reward safe behavior**
  - Positive reinforcement motivates Workers to stay vigilant and proactive.

- **Include Workers in incident investigations**
  - Involving staff in analyzing near misses and incidents helps them understand root causes and Preventive Actions.

Increasing Awareness and Responsibility Among Workers in Small Enterprises

Awareness and responsibility are closely linked in OH&S management. Awareness ensures that Workers understand hazards, safe procedures, and emergency protocols, while responsibility ensures they act on that knowledge consistently. In small businesses, where supervision may be informal and Workers multitask, cultivating both is essential for an effective OH&S Management System.

Key Practices to Increase Awareness and Responsibility

- **Provide structured induction and refresher training**
  - Introduce new **Workers** to hazards, controls, emergency procedures, and safety expectations.
  - Conduct periodic refreshers to reinforce knowledge and update Workers on new hazards or procedures.
- **Use visual aids and signage**
  - Clear signage, hazard labels, and visual reminders reinforce safe practices in the workplace.
- **Clarify roles and responsibilities**
  - Ensure every employee knows their duties, accountability, and how their actions affect workplace safety.
- **Promote direct communication channels**
  - Encourage Workers to report hazards, unsafe conditions, and incidents without fear of blame.
- **Integrate safety into daily operations**
  - Link awareness and responsibility to actual tasks so Workers understand the practical impact of safety on their work.
- **Monitor and provide feedback**
  - Observe work practices, provide constructive feedback, and recognize safe behavior to reinforce accountability.

Benefits of Participation and Responsibility

- Improves based on based on risk assessment methods defined by the organization in line with ISO 45001:2018 Clause 6.1.2" to align with risk-based thinking. and risk control.
- Enhances compliance with ISO 45001 and regulatory requirements.
- Builds a proactive safety culture where Workers take ownership of their actions.
- Reduces workplace incidents and near misses.

Table: Methods to Promote Participation, Awareness, and Responsibility in Small Enterprises

Method	Purpose	Implementation Tips
Risk Assessment Involvement	Identify hazards practically	Include Workers in task analysis and hazard review
Safety Committees / Working Groups	Discuss and solve safety issues	Small groups with management + Workers, meet regularly
Suggestion Systems	Encourage improvement ideas	Use physical or digital suggestion boxes, review submissions
Toolbox Talks / Safety Meetings	Reinforce safety knowledge	Short, focused meetings covering recent hazards or updates
Recognition & Rewards	Motivate proactive behavior	Highlight safe practices, reward suggestions or compliance
Incident Participation	Learn from near misses or accidents	Include Workers in root cause analysis and corrective action planning
Signage / Visual Aids	Increase awareness of hazards	Place clear labels, warning signs, and reminders at relevant locations
Defined Roles & Responsibilities	Ensure accountability	Document and communicate each employee's OH&S duties
Feedback & Monitoring	Maintain responsibility	Observe practices, provide feedback, and correct unsafe behavior

Conclusion

Encouraging employee participation and fostering awareness and responsibility are essential for building a **strong OH&S culture** in small enterprises. Participation empowers Workers to contribute insights, follow safe procedures, and take ownership of safety. Awareness ensures that Workers understand hazards, controls, and emergency actions. Together, these practices reduce incidents, enhance compliance, strengthen safety culture, and create a resilient, engaged workforce capable of maintaining safe operations under ISO 45001:2018 standards.

Author's Personal Note

Employee participation, awareness, and responsibility are the heart of any effective OH&S Management System, especially in small enterprises where operations are close-knit and staff often perform multiple roles. A policy or procedure alone cannot create a safe workplace; it is the active engagement of **Workers** that turns rules into practice and knowledge into safe behavior.

This chapter highlights that involving Workers in risk assessment, decision-making, and safety improvement not only strengthens compliance with ISO 45001:2018 but also fosters a culture where safety is shared, valued, and consistently applied. Encouraging participation, building awareness, and defining responsibilities empowers Workers to act proactively, recognize hazards, and support continuous improvement.

For small businesses, these practices are practical, achievable, and highly effective. Even simple, consistent efforts—like toolbox talks, suggestion systems, and clear role definitions—can significantly reduce incidents, improve operational safety, and cultivate a workforce that takes ownership of health and safety.

Kind regards,
Dr. Mohamed-Ali Ibrahim
Vienna, Austria

Congratulations on progressing through the chapters on **training, competency, and employee engagement in OH&S**. By understanding the importance of structured training programs, ongoing skill development, and active employee participation, you are equipping your small enterprise with the tools to create a safe, compliant, and resilient workplace.

Remember, ISO 45001:2018 compliance is not just about documentation or audits—it is about cultivating a culture where safety is understood, practiced, and continuously improved. Engaged and competent Workers are the foundation of this culture. Their awareness, involvement, and accountability transform policies into action and hazards into controlled risks.

Small enterprises may face unique challenges, such as limited resources, overlapping roles, or rapidly changing tasks. Yet, these challenges also create opportunities: closer communication, direct influence on daily operations, and faster implementation of safety improvements. Use the knowledge and tools from this book to build practical, ongoing, and effective OH&S practices.

Stay proactive, keep learning, and involve your Workers in creating a safer workplace. Every small step taken consistently leads to safer work environments, stronger compliance, and a thriving business.

Kind regards,
Dr. Mohamed-Ali Ibrahim
Vienna, Austria

PART 3

Practical Strategies for Small Business OH&S Compliance

A Step-by-Step Guide to Maintaining ISO 45001:2018 Compliance in Daily Operations

CHAPTER 11

Building a Strong Foundation for OH&S Compliance

A strong foundation for OH&S compliance begins with understanding that occupational health and safety is not only a legal or procedural matter, but a core management responsibility. In small enterprises, where owners, managers, and supervisors are often directly involved in daily operations, health and safety controls must be practical, visible, and consistently applied. This foundation is strengthened when the organization understands what ISO 45001:2018 requires, how those requirements support safer work, and why day-to-day compliance matters for business continuity as much as it does for worker protection. ISO describes ISO 45001 as an international standard for an occupational health and safety management system that helps organizations manage risks and improve OH&S performance, and it applies to organizations of all sizes.

Understanding ISO 45001:2018

ISO 45001:2018 is the international standard for occupational health and safety management systems. It sets out requirements, with guidance for use, to help an organization provide safe and healthy workplaces by preventing work-related injury and ill health and by proactively improving OH&S performance. In practical terms, this means the standard is not limited to reacting after accidents happen. Instead, it asks organizations to establish a structured system that identifies hazards, assesses risks, applies controls, checks whether those controls are working, and improves over time.

One of the most important points to understand is that ISO 45001 is a management system standard, not a list of fixed technical rules for every workplace. It does not prescribe one identical system for all organizations, and it does not state specific OH&S performance criteria that every organization must meet in the same way. ISO explains that the system should be appropriate to the organization's own needs, size, type of activity, and level of risk. For a small enterprise, this is especially important because it means compliance should be proportionate. A small business with limited processes and lower levels of risk does not need the same level of complexity as a large, high-risk organization, but it still needs an effective and controlled system.

ISO 45001 is built around a clear management logic. It addresses OH&S policy, objectives, planning, implementation, operation, auditing, review, and continual improvement. ISO also highlights key elements such as leadership commitment, worker participation, based on based on risk assessment methods defined by the organization in line with ISO 45001:2018 Clause 6.1.2" to align with risk-based thinking., risk assessment, legal and regulatory compliance, emergency planning, incident investigation, and continual improvement. This means the standard is not only about writing procedures. It is about making sure responsibility is defined, Workers are involved, risks are understood, controls are applied, and results are reviewed. In a well-functioning system, health and safety become part of normal management rather than a separate activity handled only when a problem occurs.

The standard follows the Plan-Do-Check-Act approach. In the planning stage, the organization considers its context, its risks and opportunities, and its legal and other requirements, then sets its policy and objectives. In the doing stage, it implements controls, assigns responsibilities, provides competence and awareness, and operates its OH&S processes. In the checking stage, it monitors performance, evaluates compliance, audits the system, and reviews results. In the acting stage, it takes corrective action and drives continual improvement. For small enterprises, this structure is useful because it turns OH&S into a repeatable cycle rather than a one-time effort.

Another key aspect of ISO 45001 is worker participation. ISO emphasizes that those closest to OH&S risks are often the people who understand them best, which is why worker participation is important in establishing, implementing, and maintaining the OH&S management system. The standard also requires that Workers be competent to carry out their assigned tasks safely. In small businesses, this principle is highly practical. Workers often notice unsafe conditions, weak controls, or procedural gaps before management does. When they are consulted, listened to, and trained properly, the system becomes more accurate, more realistic, and more effective.

It is also important to understand what ISO 45001 is not. ISO states that it is not a legally binding document in itself, but a voluntary management tool. At the same time, the standard requires organizations to determine and address applicable legal and other requirements related to OH&S. This distinction matters. Compliance with ISO 45001 does not replace compliance with the law. Instead, the standard gives the organization a structured framework for identifying legal duties, controlling risks, and maintaining a disciplined approach to OH&S management.

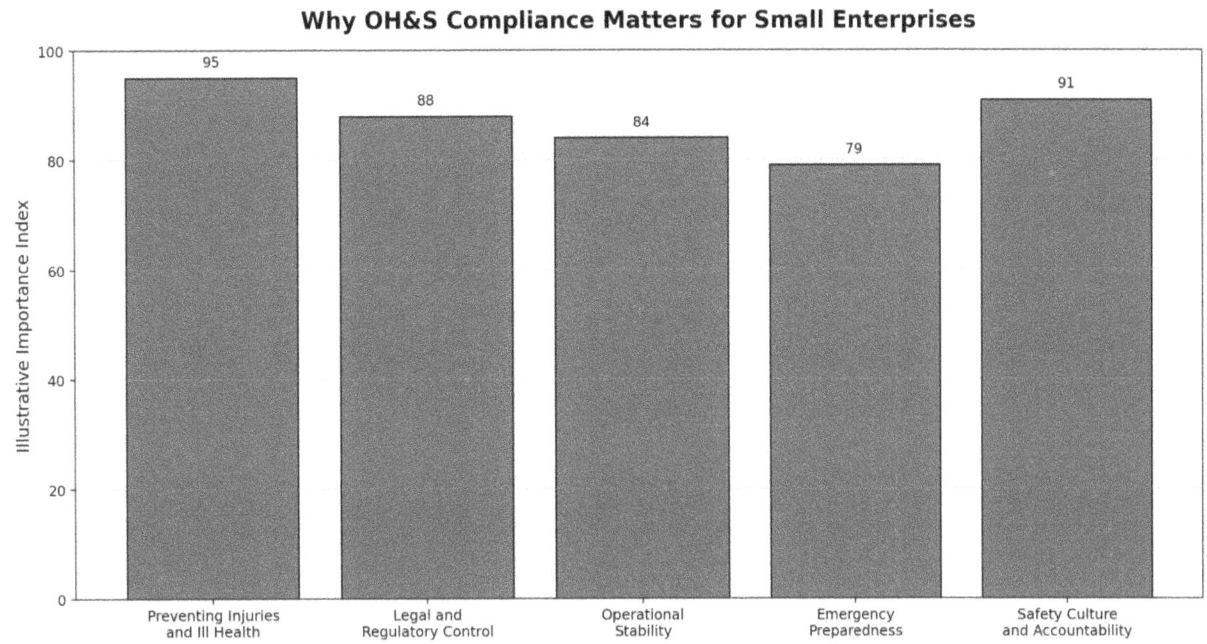

Illustrative chart for book use only (not based on survey data).

Why OH&S Compliance Matters for Small Enterprises

OH&S compliance matters greatly for small enterprises because the consequences of poor control are often more severe for smaller organizations than for larger ones. A single serious injury, unsafe event, enforcement issue, or prolonged disruption can affect staffing, customer service, reputation, productivity, and financial stability. ISO and UNIDO note that the primary focus of an OH&S management system is to prevent work-related injury, illness, or loss of life, and that beyond protecting Workers, it can also support better morale, improved services, and more stable revenue. Note that the primary focus of an OH&S management system is to prevent work-related injury, illness, or loss of life" or "prevent work-related injury, illness, or loss of life, and that beyond protecting Workers, it can also support better morale, improved services, and more stable revenue. For a small enterprise, these outcomes are not abstract benefits; they are essential conditions for sustainable operation.

Compliance also matters because small enterprises often work under conditions that can increase risk if not properly controlled. These may include multi-role staff, informal communication, rapid operational changes, reliance on a few key people, limited documentation, and close contact between Workers, visitors, contractors, and equipment. In such environments, health and safety weaknesses can become embedded in routine work unless they are actively managed. ISO 45001 helps organizations establish systematic processes that consider context, risks and opportunities, hazards, and legal requirements, rather than relying only on personal experience or informal habits.

For small enterprises, OH&S compliance is also important because it improves operational discipline. When hazards are identified, controls are applied, Workers are trained, incidents and near misses are monitored, and corrective actions are followed through, the organization becomes more stable and less reactive. ISO points out that practical first steps include identifying hazards and unsafe conditions, involving Workers in safety discussions, reviewing legal obligations, documenting procedures and controls, training Workers and supervisors, monitoring incidents and near misses, and with "and planning corrective actions and Emergency arrangements should be communicated clearly to Workers actions and preventive measures. actions and preventive measures, use "corrective actions and Emergency arrangements should be communicated clearly to Workers. These are not merely administrative tasks. They are the building blocks of daily operational control.

Another reason compliance matters is that it supports legal and regulatory readiness. ISO explains that an ISO 45001-based system helps organizations respond to regulatory compliance issues and establish operational controls to manage both OH&S risks and legal requirements. For a small business, this can reduce the likelihood of overlooked duties, inconsistent practices, or unmanaged changes that later become compliance failures. A structured OH&S approach makes it easier to assign responsibilities, keep records, review actions, and demonstrate that health and safety is being actively managed rather than passively assumed.

OH&S compliance also strengthens resilience. ISO highlights emergency preparedness and response as a key part of the standard and notes that the framework helps organizations respond more effectively to safety threats and crises. In a small enterprise, resilience is critical because disruptions can be harder to absorb. A well-managed OH&S system helps the business prepare for foreseeable emergencies, respond in a controlled way, recover more quickly, and learn from events so that the same failures are less likely to happen again.

Finally, OH&S compliance matters because it shapes culture. When leaders show commitment, Workers are involved, controls are maintained, and follow-up is consistent, safety becomes part of how the business operates. ISO notes that adopting the standard demonstrates commitment to worker health, safety, and wellbeing, while also supporting morale and retention. In small enterprises, where relationships are closer and management actions are more visible, this cultural effect can be especially strong. Workers are more likely to report problems, follow controls, and take OH&S seriously when they see that compliance is treated as a real priority in daily decision-making.

The Purpose and Benefits of an OH&S Management System

An occupational health and safety management system is a structured way for an organization to manage health and safety as part of normal business activity. Under ISO 45001:2018, it is not meant to be a separate file of procedures that sits apart from real work. It is a management framework that helps the organization identify hazards, assess risks, apply controls, review performance, and improve over time. ISO describes it as a framework to manage risks and improve OH&S performance, with requirements covering policy, objectives, planning, implementation, operation, auditing, review, and continual improvement.

The core purpose of an OH&S management system is to provide safe and healthy workplaces by preventing work-related injury and ill health. This purpose is especially important in small enterprises, where even one serious incident can disrupt operations, reduce productivity, affect staff confidence, and place pressure on already limited resources. The ISO/UNIDO practical guide for small organizations notes that the negative effects of work-related injuries, illnesses, or fatalities can be overwhelming for small businesses, which is why a systematic approach is so valuable.

A second purpose of the system is to move the organization from a reactive approach to a preventive one. Instead of acting only after an accident, complaint, or inspection finding, the organization is expected to establish a process that identifies unsafe conditions early, involves Workers, sets controls in advance, and checks whether those controls are working. ISO's current overview of the standard explains that organizations use ISO 45001 to identify hazards, assess risks and opportunities, implement controls, involve Workers, monitor incidents and near misses, and review and improve the system regularly.

The benefits of an OH&S management system extend beyond accident prevention. ISO states that implementing ISO 45001 can help reduce workplace incidents, demonstrate commitment to worker health and safety, support legal compliance, strengthen resilience through emergency preparedness, and improve OH&S performance through continual improvement. The ISO/UNIDO handbook for small organizations also highlights broader benefits such as better risk management, stronger worker engagement, improved morale, improved recruiting and retention, and a more favorable image and reputation.

For small enterprises, one of the most practical benefits is operational discipline. When responsibilities are clear, hazards are identified, controls are maintained, inspections are carried out, and problems are corrected in time, the workplace becomes more stable and less dependent on guesswork. The management system helps turn safety from an occasional concern into a managed routine. That is why ISO 45001 is valuable not only for high-risk industries, but for organizations of all sizes and sectors.

Aspect	What It Means in Practice	Main Benefit for a Small Enterprise
Protection of Workers	The organization identifies hazards, controls risks, and works to prevent injury and ill health	Fewer incidents, safer working conditions, and stronger worker confidence
Structured management	OH&S is handled through policy, planning, operation, review, and	Better control, clearer responsibilities, and more consistent decisions

	improvement rather than informal reactions	
Legal compliance	The system helps the organization identify and address applicable OH&S obligations	Reduced risk of missed duties, weak controls, and compliance failures
Emergency readiness	The organization plans and prepares for potential emergencies before they occur	Faster, more organized response and reduced disruption
Continual improvement	Performance is checked, problems are corrected, and controls are improved over time	A stronger and more reliable OH&S system as the business develops

This table summarizes the main role of an OH&S management system as described by ISO 45001 and the ISO/UNIDO small-organization guidance.

Understanding Legal and Other Requirements

In ISO 45001, legal and other requirements are a central part of compliance. ISO's current guidance lists the determination of legal and other requirements as one of the key requirements of the standard, and the standard page itself highlights legal and regulatory compliance as a key element of the OH&S management system. This means an organization is not expected to manage OH&S only through internal preference. It must also understand the external obligations that apply to its work.

Legal requirements are the OH&S laws, regulations, and statutory duties that the organization has to comply with in the places where it operates. These may relate to workplace safety, hazardous substances, machinery, emergency arrangements, training, reporting, consultation, working conditions, contractor control, or other regulated matters, depending on the nature of the business. ISO emphasizes that the standard helps organizations comply with OH&S regulations, but it does not replace national law; ISO standards are voluntary, while legal requirements take precedence.

Other requirements are additional OH&S obligations that the organization has to comply with or chooses to comply with. ISO's terminology database and browsing platform use this exact distinction: legal requirements are those the organization has to comply with, while other requirements are those it has to or chooses to comply with. In practice, these can arise from customer conditions, supply-chain expectations, corporate rules, insurer conditions, site rules, tender requirements, or internal commitments that the organization has formally adopted.

Understanding these requirements is not a one-time exercise. Small enterprises often change sites, equipment, contractors, services, or work methods, and each change can affect what obligations apply. That is why the organization must not only identify relevant requirements, but also keep them current, communicate them where necessary, and reflect them in operational controls. If a requirement exists but is not translated into real practice, the organization remains exposed to compliance failure. ISO's practical first steps include reviewing legal OH&S obligations, documenting procedures and controls, training Workers and supervisors, and planning corrective and Preventive Emergency arrangements should be communicated clearly to Workers.

In daily business terms, this means a small enterprise should know which OH&S obligations apply, where that information is kept, who is responsible for maintaining it, how changes are identified, and how compliance is checked in actual work. A requirement has little value if it remains unknown to the people who supervise or perform the task. For this reason, legal and other requirements should be linked to hazards, tasks, procedures, inspections, and management review so that they become part of working practice rather than isolated paperwork.

Using the Plan-Do-Check-Act Approach in Daily Business Activities

The Plan-Do-Check-Act approach is the operating logic behind ISO 45001. ISO states that the standard uses the Plan-Do-Check-Act methodology to systematically manage health and safety risks, and the ISO/UNIDO guide for small organizations explains the cycle in simple terms: plan what needs to be done to prevent workplace ill health and injury, do what was planned, check how well it is working, and act to fix problems and improve effectiveness.

In daily business activities, the **Plan** stage means deciding in advance how OH&S risks will be controlled. This includes understanding the business context, identifying hazards, considering risks and opportunities, determining legal and other requirements, setting OH&S objectives, and deciding what controls, resources, responsibilities, and timeframes are needed. In a small enterprise, planning may be simple, but it still needs to be clear. Unsafe work should not be left to assumptions or personal habits.

The **Do** stage means putting the planned arrangements into practice. This includes providing resources, ensuring competence and awareness, communicating requirements, controlling documented information where needed, implementing operational controls, and preparing for emergencies. In everyday terms, this is the stage where safe work practices are followed, supervision takes place, PPE is used where required, contractors are controlled, and Workers know what is expected of them. The system becomes real only when the planned controls are actively used in work.

The **Check** stage means verifying whether the system is actually working. ISO links this stage to monitoring, measurement, performance evaluation, compliance evaluation, internal audit, and review. For a small enterprise, checking may involve workplace inspections, review of incidents and near misses, follow-up of overdue actions, supervisor observations, simple performance indicators, and management discussions about whether controls are effective. This stage is essential because a control that exists on paper but fails in practice is not enough.

The **Act** stage means responding to what the organization has learned. Problems are corrected, nonconformities are addressed, causes are considered, actions are taken to prevent recurrence, and the system is improved. This is where the organization strengthens the weak parts of its OH&S arrangements. ISO presents continual improvement as a key outcome of the management system, and the PDCA model is the mechanism that keeps improvement active rather than accidental.

For small enterprises, the real value of PDCA is that it converts OH&S from a one-time setup into a living management process. Planning without doing creates paperwork. Doing without checking creates false confidence. Checking without acting leaves known problems unresolved. Acting without planning leads to inconsistency. The strength of the model is that each part supports the others, creating a practical cycle that can be repeated in routine operations, during change, after incidents, and during review.

A Personal Note from the Author

In this chapter, I wanted to build the reader's understanding of why occupational health and safety compliance must begin with a strong foundation. Small enterprises often face real operational pressures, limited resources, and fast-moving daily demands, which can make safety seem difficult to manage in a structured way. For that reason, I focused on explaining ISO 45001:2018 in practical terms and showing why OH&S compliance is not only a formal requirement, but a necessary part of responsible business management.

I also wanted to make clear that compliance is not just about avoiding problems. It is about protecting Workers, strengthening daily operations, improving consistency, and helping small businesses work with greater confidence and control. My intention in this chapter is to help readers see that a solid OH&S foundation is the starting point for every other part of the system. When the basics are understood correctly, the rest of the journey becomes more practical, more effective, and more meaningful.

CHAPTER 12

Leading Safety with Clear Policy and Smart Planning

Chapter Two focuses on the part of ISO 45001:2018 that turns health and safety from a general intention into a managed business direction. ISO presents **leadership and worker participation** as key elements of the standard, and it states that the standard establishes criteria for an **OH&S policy**, objectives, planning, implementation, operation, auditing, review, and continual improvement. For small enterprises, this means safety leadership is not separate from management. It must be visible in how the business is directed, how priorities are set, and how daily work is controlled.

Creating an Effective OH&S Policy

An OH&S policy is the formal statement that explains the organization's direction and intentions for occupational health and safety. Under ISO 45001, the policy is not just a display document for a wall or an audit file. It is part of the management system framework and should guide decisions, objectives, operational control, and improvement activities. ISO's public overview confirms that the standard establishes criteria for an OH&S policy, while its practical small-organization handbook explains that the standard is meant to help small organizations establish an OH&S management system that works for the organization and supports safer work.

In practical terms, an effective OH&S policy gives the organization a clear safety direction that people can understand and apply. In a small enterprise, this is especially important because communication is often informal and managers are closely involved in operations. If the policy is vague, overly technical, or disconnected from real work, it will not influence daily practice. A useful policy should therefore reflect the nature of the business, the type of work performed, the risks that arise from that work, and the organization's intention to manage OH&S in a systematic way. This practical interpretation fits ISO's emphasis that ISO 45001 applies to organizations of all sizes and that systems should be proportionate to the organization's context, risk profile, and complexity.

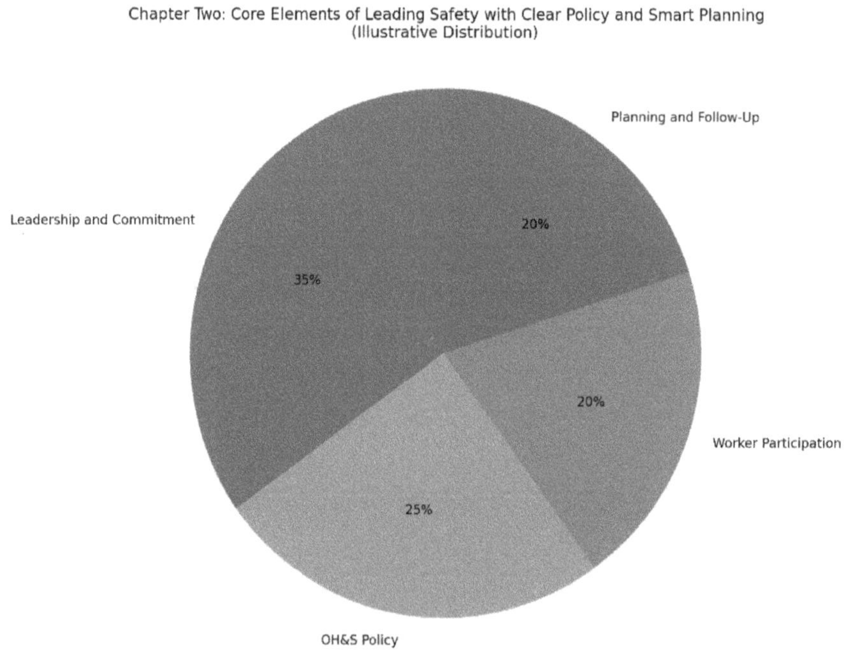

An effective policy also matters because it gives structure to planning. Once the business has declared its OH&S direction, that direction should be reflected in objectives, responsibilities, procedures, supervision, and follow-up. A policy that says safety is important but is not connected to training, hazard control, reporting, inspections, or corrective action has little value. ISO's current explanation of the standard links policy with objectives, planning, implementation, operation, review, and continual improvement, which shows that the policy is intended to shape the whole system rather than stand alone.

For small enterprises, the strongest OH&S policies are usually the clearest ones. They do not try to sound legalistic or overly broad. Instead, they communicate management's commitment in a form that Workers, supervisors, and contractors can understand. When the policy is realistic and connected to actual workplace conditions, it becomes easier to communicate expectations, explain priorities, and support consistent action. This reflects the ISO/UNIDO guidance that small organizations should develop a system that works for the organization rather than copy unnecessary complexity.

Policy Element	Why It Matters	What It Looks Like in a Small Enterprise
Clear direction	Gives the organization a defined OH&S position	Management states what safety means in practice for the business
Practical relevance	Prevents the policy from becoming disconnected from work	The policy reflects real tasks, hazards, and working conditions
Link to planning	Connects intentions with objectives and controls	The policy supports actions, responsibilities, and follow-up
Communication value	Helps Workers understand expectations	The wording is simple enough to be explained and applied
Basis for improvement	Supports review and continual improvement	Management can compare daily practice against stated commitments

This table summarizes how an OH&S policy becomes effective in practical management terms while remaining aligned with ISO 45001's role for policy within the wider OH&S management system.

Demonstrating Leadership and Commitment

Leadership and commitment are central to ISO 45001 because safety performance depends heavily on management behavior. ISO's public guidance states that the success of the OH&S management system depends on leadership, commitment, and participation from all levels and functions of the organization. Its 2018 launch material adds that top management must take a **visible, directing role**, be actively involved in implementation, and ensure the OH&S management system is integrated into overall business processes. This is especially significant for small enterprises, where leaders are often close to operations and their actions are seen directly by Workers every day.

Demonstrating leadership in this context means more than approving documents or delegating safety to one person. It means showing, through decisions and actions, that OH&S matters in the way work is planned and managed. In practice, this includes setting expectations, assigning responsibilities, providing resources, removing obstacles to safe work, reviewing performance, and following through when weaknesses are identified. ISO's current guidance describes leadership and worker participation as key requirements, and its public materials emphasize that OH&S should be integrated into everyday operations and decision-making.

Leadership is also closely linked to culture. ISO's launch coverage explains that senior leaders should promote a positive culture and communicate not only what needs to be done, but why it is important. In small businesses, that message has strong practical effects. When owners and managers take shortcuts, ignore unsafe conditions, or treat safety as secondary, Workers usually notice quickly. By contrast, when leaders ask about hazards, support reporting, respond to problems, and make time for prevention, Workers are more likely to participate honestly and take controls seriously.

Another important part of leadership under ISO 45001 is worker participation. ISO's public explanation highlights involving Workers in safety discussions and identifies worker participation as a key element of the standard. The launch material further states that the standard places greater emphasis on Workers actively participating in the development, planning, implementation, and continual improvement of the OH&S management system. For small enterprises, this is not an extra activity. It is often the most direct way to identify hazards early, test whether controls are realistic, and keep the system grounded in actual work.

Leadership and commitment therefore become visible in daily business activities. They are seen in how managers respond to hazards, how they balance production and safety, whether they maintain controls during busy periods, whether they involve Workers in decisions, and whether they correct problems promptly. ISO's practical guide for small organizations notes that the effects of injuries, ill health, or fatalities can be overwhelming for smaller organizations, which is why strong leadership is not simply a formal requirement; it is a practical necessity for continuity and responsible management.

Defining Roles, Responsibilities, and Authorities

In ISO 45001:2018, roles, responsibilities, and authorities are not treated as a formality. They are part of the core leadership structure of the OH&S management system, and the standard specifically includes **organizational roles, responsibilities and authorities** as a distinct requirement. ISO's current public material also states that the standard establishes criteria for policy, objectives, planning, implementation, operation, auditing, and review, which means these roles must support the full system, not only isolated safety tasks.

In practical terms, a **role** is the function a person performs in the organization, a **responsibility** is what that person is expected to do, and an **authority** is the power given to make decisions, require action, or stop unsafe work when necessary. These three elements must work together. If someone is told to manage safety but is given no authority, the system becomes weak. If authority is present but responsibility is unclear, actions become inconsistent. If both exist but are not communicated, gaps and overlaps appear in daily operations. This is why ISO management system practice places strong emphasis on assigning and communicating responsibility and authority for relevant roles.

For small enterprises, this area is especially important because one person often performs several functions. An owner may also act as manager, purchaser, and site decision-maker. A supervisor may oversee production, training, and workplace checks. That is acceptable, but the OH&S responsibilities still need to be clear. Someone must know who is responsible for inspections, who follows up corrective actions, who communicates safety requirements to Workers and contractors, who checks legal updates, and who reports OH&S performance to top management. When this clarity is missing, the business often relies on assumptions rather than control.

Clear allocation of roles also supports accountability. When Workers know who is responsible for decisions, reporting, and follow-up, problems are more likely to be raised early and handled properly. In a mature OH&S system, responsibility is visible in daily practice. Managers review performance, supervisors enforce controls, Workers follow procedures and report concerns, and leadership ensures the system remains integrated into business processes. ISO's 2018 launch material stresses that top management must take a visible, directing role and be actively involved in integrating the OH&S management system into the organization's overall business processes.

Encouraging Worker Consultation and Participation

Worker consultation and participation is one of the clearest distinguishing features of ISO 45001. It is not optional guidance. ISO's official Q&A states that participation and consultation of Workers is a requirement of ISO 45001, and top management must ensure that it takes place if the organization wishes to claim conformity with the standard. ISO's public pages and launch material also emphasize worker participation as a key element of the standard.

Consultation means Workers are asked for their views before decisions are made on matters that affect health and safety. Participation goes further: Workers actively take part in the development, planning, implementation, and continual improvement of the OH&S management system. ISO's launch coverage states that the standard places greater emphasis on Workers actively participating in these areas, because people closest to the work often understand the risks, obstacles, and impractical controls better than anyone else.

For small enterprises, consultation and participation are highly practical. Workers may notice unsafe conditions, recurring shortcuts, equipment defects, workload problems, or weak supervision long before these issues appear in records or formal reviews. If management only gives instructions without listening, the system becomes incomplete. By contrast, when Workers are consulted about procedures, task changes, incident causes, and control measures, the organization gains more accurate information and better chances of preventing harm. This supports the wider ISO 45001 approach, which links worker participation with better OH&S practices and stronger system performance.

Encouraging consultation and participation also helps build trust. Workers are more likely to report hazards, near misses, and unsafe behavior when they believe management will listen and respond fairly. In small workplaces, where daily relationships are close and management behavior is highly visible, this effect can be especially strong. A positive reporting culture does not happen by chance. It grows when managers make time for discussion, explain why safety decisions matter, and treat worker input as part of normal management rather than as resistance or complaint.

Identifying Hazards, Risks, Opportunities, and Legal and Other Requirements

This section should be handled carefully in ISO terminology. In practical business language, many people speak about **compliance obligations**, but in ISO 45001 the exact expression is **legal requirements and other requirements**. The standard also addresses **based on based on risk assessment methods defined by the**

organization in line with ISO 45001:2018 Clause 6.1.2" to align with risk-based thinking., **assessment of OH&S risks and other risks**, **assessment of OH&S opportunities and other opportunities**, and **determination of legal requirements and other requirements**. These elements sit at the heart of planning because an organization cannot control what it has not identified.

A hazard is any work-related situation, activity, or process that can harm people. ISO's current OH&S guidance explains that hazards can range from physical dangers such as frayed wiring or work at height to chemical exposure, lack of adequate resources, or unmanageable workloads. Hazards that are not eliminated or controlled create OH&S risks. This distinction matters. The hazard is the source or situation with the potential to cause harm, while the risk concerns the chance and consequence of that harm arising under actual working conditions.

ISO 45001 does not stop at risk alone. It also requires organizations to consider opportunities. ISO's public explanation states that organizations use ISO 45001 by identifying hazards, assessing risks and opportunities for improvement, implementing controls, involving Workers, and reviewing the system regularly. In a small enterprise, opportunities may include better task design, clearer supervision, safer equipment choices, stronger communication, improved training, or more reliable contractor control. The purpose is not only to prevent failure, but also to strengthen the system before problems occur.

Legal and other requirements complete this planning picture. Legal requirements are the OH&S duties that apply under law or regulation. Other requirements are additional obligations the organization has to comply with or chooses to comply with, such as customer requirements, site rules, insurer conditions, contractual demands, or internal commitments. For small enterprises, these requirements should not remain as separate reference material. They need to be translated into procedures, supervision, training, inspections, and management follow-up. Otherwise, compliance exists only on paper.

Setting OH&S Objectives and Planning Actions to Achieve Them

OH&S objectives are the practical results the organization intends to achieve through its OH&S management system. ISO 45001 includes **OH&S objectives and**

planning to achieve them as a specific requirement, showing that health and safety management must move beyond general intention and into measurable direction. The standard's public overview confirms that objectives and planning are part of the formal criteria of ISO 45001, and that the system is built on the Plan-Do-Check-Act approach.

In a small enterprise, objectives should be realistic, relevant, and connected to actual risks and controls. An objective might relate to reducing manual-handling injuries, improving completion of workplace inspections, strengthening contractor control, closing overdue corrective actions, or increasing near-miss reporting. The important point is that the objective should support the OH&S policy and respond to the organization's real conditions rather than exist as a vague statement of good intention. When objectives are clear, management can assign responsibility, provide resources, set timeframes, and check whether progress is being made. That is how planning becomes operational rather than symbolic.

Planning actions to achieve OH&S objectives means deciding in advance what will be done, who will do it, what resources are needed, when action should be completed, and how results will be evaluated. This logic aligns with the small-organization PDCA explanation published by ISO and UNIDO: plan what needs to be done to prevent ill health and injury, do what was planned, check how well it is working, and act to fix problems and improve effectiveness. In everyday practice, this means objectives should not be separated from inspections, supervision, training, maintenance, communication, and review.

Well-planned objectives also help prevent drift in the system. Without them, a business may claim that safety is important while repeating the same weaknesses month after month. Objectives create direction, while action planning creates accountability. For small enterprises, this is especially valuable because resources are limited and attention is often divided across many priorities. Clear objectives help management decide what must be controlled first, what progress is expected, and how improvement will be measured over time.

Area	Main Practical Question	Why It Matters
Roles, responsibilities, and authorities	Who is responsible, and who has the authority to act?	Prevents gaps, overlap, and weak accountability

Worker consultation and participation	Are Workers being heard and actively involved?	Improves realism, trust, and early problem detection
Hazards, risks, opportunities, and legal/other requirements	What can cause harm, what can improve control, and what obligations apply?	Supports sound planning and stronger compliance
Objectives and action planning	What result is the organization trying to achieve, and how will it get there?	Converts OH&S intentions into measurable action

This table summarizes how these four areas work together to support a practical and functioning OH&S management system in line with ISO 45001:2018.

A Personal Note from the Author

In this chapter, I wanted to emphasize a truth that is often overlooked in small enterprises: occupational health and safety does not become effective simply because a policy exists or responsibilities are written down. It becomes effective when leadership is visible, when responsibilities are clearly understood, when Workers are genuinely involved, and when planning is based on real workplace conditions rather than assumptions. A strong OH&S foundation is built through daily management attention, practical communication, and a willingness to act before problems become incidents.

I also believe that small businesses should not view ISO 45001:2018 as an unnecessary burden or as a system designed only for large organizations. When understood properly, it provides a practical structure for safer work, clearer accountability, and better decision-making. My intention in this chapter is to show that strong leadership and smart planning are not abstract management ideas. They are the starting point of a safer, more disciplined, and more responsible workplace.

CHAPTER 13

Turning OH&S Requirements into Everyday Practice

Chapter Three moves the discussion from policy and planning into practical application. In ISO 45001:2018, this part of the system sits mainly within the **Support** clauses, where the organization must provide the conditions that allow OH&S controls to work in reality. ISO's public overview identifies **training, awareness and competence** as key requirements of the standard, and the ISO Online Browsing Platform lists Clause 7 as including **resources, competence, awareness, communication, and documented information**. This means a business cannot expect safe performance simply by setting objectives or writing procedures. It must also provide the people, time, tools, knowledge, and support needed to carry those requirements into daily work.

Providing the Resources Needed for OH&S

Providing resources for OH&S means making sure the organization has what is necessary to establish, implement, maintain, and improve the occupational health and safety management system. ISO's browsing platform identifies **7.1 Resources** as a specific requirement, and ISO's small-organization handbook explains that small organizations implementing ISO 45001 often need to address resources alongside competence and documented information as part of building a functioning system. In practical terms, resources are not limited to money. They include people, time, equipment, information, protective measures, access to expertise, maintenance support, and the managerial attention needed to keep safety controls active.

For small enterprises, resource provision is one of the most important practical tests of leadership. A company may say that safety matters, but if supervisors have no time to inspect work, if equipment is not maintained, if training is delayed, if corrective actions stay open for months, or if Workers are expected to manage hazards without proper tools or support, the OH&S system becomes weak. ISO's hazards and risks guidance specifically includes **providing necessary resources for**

OH&S management as a core organizational action. This shows that resource provision is not optional support around the edges of the system; it is part of how risks are actually controlled.

Resources also need to match the size and nature of the business. ISO states that ISO 45001 applies to organizations of all sizes and sectors, and its briefing materials explain that the system should be appropriate to the organization's context, activities, and risk level. For a small enterprise, this means the resource base may be simpler than in a large organization, but it still has to be sufficient. A small warehouse may not need a large safety department, but it still needs clear supervision, maintained equipment, emergency arrangements, competent Workers, and enough time for safety checks and corrective follow-up. Proportionate does not mean minimal to the point of ineffectiveness. It means suitable to the actual work and its risks.

Another important point is that resources should support both prevention and improvement. They are needed not only for normal work, but also for inspections, incident investigations, communication, emergency readiness, and continual improvement. ISO's public explanation of practical first steps includes documenting procedures and controls, training Workers and supervisors, monitoring incidents and near misses, and planning corrective and Preventive Emergency arrangements should be communicated clearly to Workers. All of these activities depend on resources. If the organization provides resources only for production and not for prevention, then OH&S performance will remain unstable.

In daily business terms, providing resources for OH&S means asking practical questions. Are Workers given enough time to perform work safely rather than rush? Are the tools, machines, and protective measures suitable and maintained? Is competent supervision available? Are hazards reviewed when operations change? Are actions from audits, incidents, or inspections actually funded and followed through? These are the real signs that the organization is supporting OH&S instead of treating it as a secondary activity. In small enterprises, where budgets and staffing are often limited, this kind of disciplined prioritization is essential.

Building Competence, Awareness, and Training

ISO 45001 treats competence, awareness, and training as closely connected, but they are not exactly the same thing. ISO's public summary of the standard lists **training, awareness and competence** together as key requirements, while the browsing platform separates them into **7.2 Competence** and **7.3 Awareness**. This distinction matters. Training is one way to build competence, but competence is the ability to perform work safely and effectively, while awareness is the person's understanding of OH&S policy, hazards, controls, responsibilities, and the consequences of not following requirements. A business may deliver training sessions, but if Workers still do not understand the risks or cannot perform the task safely, the requirement has not truly been met.

Competence is fundamental because the OH&S system depends on people carrying out tasks correctly under real working conditions. ISO's current hazards guidance identifies **ensuring competence, training and awareness** as a core element of effective OH&S management. In practice, competence means Workers, supervisors, and others doing work under the organization's control have the necessary knowledge and ability for the activities they perform. This includes understanding safe methods of work, recognizing hazards, using controls correctly, responding to emergencies when relevant, and knowing when to stop and report unsafe conditions. Competence is therefore practical performance, not just attendance at a briefing.

Training plays an important role because it helps close the gap between what a person currently knows and what the job requires. ISO's practical first-step guidance specifically mentions **training Workers and supervisors** as part of building an effective OH&S management system. In a small enterprise, training may take different forms depending on the work: induction for new Workers, task-specific instruction, refresher training, toolbox talks, supervisor coaching, contractor briefings, or emergency drills. What matters is that the training is relevant to actual risks and is delivered in a way people can understand and apply. Training that is generic, rushed, or disconnected from daily tasks rarely produces reliable safety performance.

Awareness goes beyond technical instruction. It means people understand the OH&S direction of the business and their part in it. Workers should be aware of the hazards related to their work, the controls that must be followed, the reporting channels available to them, and the importance of complying with procedures and instructions. Awareness is also cultural. When Workers understand why controls exist and how their actions affect others, they are more likely to act consistently and report problems early. ISO's launch and explanatory materials repeatedly connect worker participation with stronger OH&S performance, which shows that awareness is not passive knowledge; it supports informed involvement in the system.

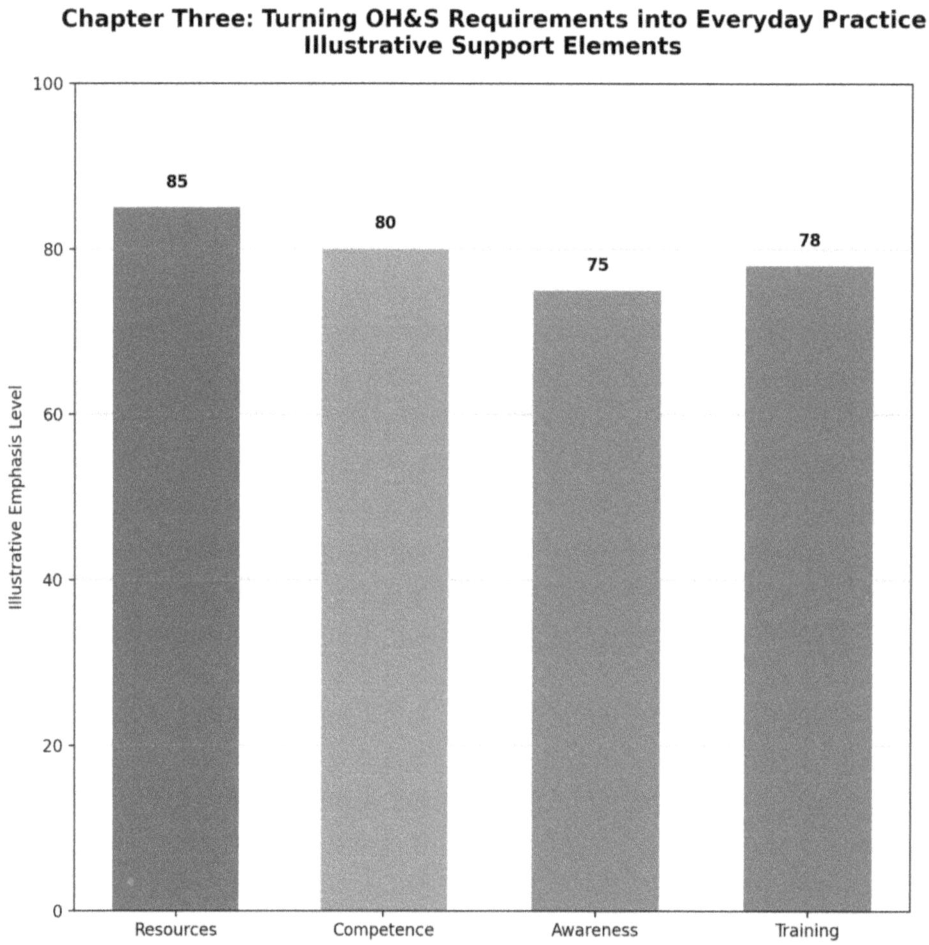

For small enterprises, the relationship between competence, awareness, and training is especially important because teams are often small and roles can overlap. A person may operate equipment, handle materials, supervise others, and respond to minor incidents in the same working day. That means competence needs to reflect the real demands of the role, not a narrow job description. It also means awareness should be reinforced regularly through daily communication, supervision, and practical reminders, not left only to induction day. ISO's materials for smaller organizations emphasize implementation approaches that are proportionate and workable, which supports using simple but effective methods to maintain competence and awareness over time.

An important risk for many businesses is confusing training records with real competence. A signed attendance sheet may show that a session happened, but it does not by itself show that the person understood the content, can apply it safely, or will act correctly under pressure. That is why competence should be supported by observation, supervision, questioning, feedback, and, where appropriate, refresher training. In practical OH&S management, the real evidence of competence is seen in safe behavior, correct use of controls, early recognition of hazards, and consistent adherence to procedures during actual work. ISO's performance-focused structure supports this practical view because the standard is designed to improve OH&S performance, not just produce documentation.

Building competence and awareness also strengthens the wider management system. Competent and aware Workers are more likely to identify hazards, participate in safety discussions, follow safe work practices, respond properly to changes, and contribute to continual improvement. This is one reason ISO places competence and awareness in the Support section rather than treating them as isolated human resource issues. They support planning, operation, emergency preparedness, performance evaluation, and improvement across the whole OH&S management system. For small enterprises, this makes them a strategic requirement, not just a training activity.

Taken together, providing resources and building competence, awareness, and training are what make OH&S requirements workable in everyday practice. Resources provide the means. Competence provides the capability. Awareness provides understanding and attention. Training helps develop and refresh both competence and awareness. When these elements are present and actively maintained, the organization is much better placed to turn OH&S requirements into reliable daily behavior rather than occasional compliance effort.

Strengthening Internal and External Communication

Communication is one of the practical support functions that makes an OH&S management system work in real life rather than only on paper. ISO 45001 presents implementation and operation as core parts of the system, and the ISO small-organization guidance explains that the OH&S management system must be integrated into business processes and day-to-day work. In practice, that cannot happen without clear communication inside the organization and, where relevant, with parties outside it.

Internal communication means the movement of OH&S information between owners, managers, supervisors, Workers, and any others working under the organization's control. It should help people understand hazards, controls, responsibilities, changes in work methods, emergency arrangements, incident Observations and recommendations must be documented; implementation is per organization discretion., and corrective actions. In a small enterprise, internal communication is especially important because work is often fast-moving, roles overlap, and informal instructions can easily replace controlled information. When communication is weak, people may not know what has changed, what control is required, or who is responsible for acting.

External communication concerns OH&S-related information that must be shared with parties outside the organization, such as contractors, visitors, suppliers, customers, landlords, insurers, regulators, or emergency services, depending on the nature of the business. For small enterprises, this often includes communicating site rules, contractor expectations, emergency arrangements, hazard information, and relevant safety conditions before work begins. Good external communication reduces the chance that outside parties will introduce unmanaged risks into the workplace or misunderstand the organization's safety requirements.

Strong OH&S communication is not just about sending messages. It must also allow information to move back to management. That is why communication supports worker participation, hazard reporting, incident follow-up, and continual improvement. If Workers or contractors cannot raise concerns clearly and promptly, the organization loses one of its most useful sources of risk information. In small businesses, where leaders are close to operations, communication should be direct, timely, and practical enough to support immediate action when needed.

A practical communication process usually includes:

- who needs the information

- what information must be shared
- when it must be communicated
- how it will be communicated
- how responses, questions, or concerns will be received and acted on

Creating and Controlling Documented Information

Documented information is the recorded information the organization needs in order to run the OH&S management system effectively and show that important controls are in place. ISO's small-organization guide identifies documented information as one of the common management-system elements that organizations implementing ISO 45001 need to manage, and the harmonized ISO management-system structure requires documented information to be created, updated, and controlled so it is available where needed and protected from loss, misuse, or unintended change.

In practical terms, documented information can include OH&S policies, procedures, safe work instructions, inspection forms, training records, maintenance records, incident reports, risk assessment outputs, emergency arrangements, and corrective-action records. The exact extent of documentation should be appropriate to the size, activities, and complexity of the business. A small enterprise does not need unnecessary paperwork, but it does need enough documented information to communicate requirements clearly, keep work consistent, support accountability, and retain evidence of what has been done.

Control of documented information is just as important as creating it. A document that is outdated, missing, unreadable, or unavailable at the point of use can create confusion and lead to unsafe decisions. That is why documented information should be identified clearly, reviewed and approved appropriately, kept current, accessible where needed, and protected through suitable storage, version control, retention, and disposal arrangements. This is a standard management-system discipline and supports OH&S consistency in daily work.

For small enterprises, the real purpose of documented information is control, not bureaucracy. A short, clear instruction that Workers actually use is more valuable than a large file no one consults. The best documented information supports daily work, helps supervisors verify what should happen, and provides a reliable basis for training, review, and corrective action. When documentation is practical and controlled, it strengthens the whole OH&S system.

Developing Safe Work Practices

Safe work practices are the practical operating methods used to control hazards during routine and non-routine work. ISO 45001 is built around based on based on risk assessment methods defined by the organization in line with ISO 45001:2018 Clause 6.1.2" to align with risk-based thinking., risk assessment, legal and regulatory compliance, implementation, operation, emergency planning, and continual improvement. In that framework, safe work practices are one of the clearest ways an organization translates OH&S requirements into everyday action. They help turn general system requirements into specific instructions for how work should actually be done.

A safe work practice should reflect the real conditions of the task. It should consider the work steps involved, the hazards that may arise, the controls that must be applied, the people responsible, the equipment or PPE required, and what must happen if conditions change or something goes wrong. For small enterprises, this is especially important because work is often done by small teams, under time pressure, and in changing environments. A practice that is too generic or disconnected from the task will not control risk effectively.

Safe work practices also support consistency. They help reduce reliance on memory, habit, or personal interpretation by making expectations clearer. This is useful for new Workers, temporary staff, contractors, and supervisors, and it becomes even more important when equipment changes, tasks are infrequent, or risks are higher than normal. When safe work practices are reviewed after incidents, inspections, or operational changes, they also become an important tool for continual improvement.

A good safe work practice usually makes these points clear:

- the task or activity being controlled
- the main hazards and risk points
- the required controls and protective measures
- who is authorized or competent to do the work
- what to report, stop, or escalate if conditions become unsafe

Area	Main Purpose	Practical Value in a Small Enterprise
Internal and external communication	Ensures OH&S information is shared with the right people at the right time	Improves coordination, reporting, contractor control, and response to change
Documented information	Records and controls the information needed to run the OH&S system effectively	Supports consistency, accountability, evidence, and reliable follow-up
Safe work practices	Converts OH&S requirements into task-level operating controls	Helps Workers perform jobs safely and consistently under real conditions

Together, communication, documented information, and safe work practices form a practical bridge between OH&S planning and OH&S performance. Communication keeps people informed and engaged, documented information keeps requirements clear and controlled, and safe work practices turn those requirements into repeatable action at the workplace.

Applying the Hierarchy of Controls

The hierarchy of controls is one of the most practical concepts in OH&S management because it helps an organization choose controls in the right order, based on effectiveness. ISO's current occupational safety guidance describes it as a simple method that prioritizes solutions according to how effective they are, starting with the strongest way to remove risk and then working downward. The five levels are elimination, substitution, engineering controls, administrative controls, and personal protective equipment. ISO also explains that PPE should be used only when higher-level controls are not enough.

In practical terms, applying the hierarchy means the organization should first ask whether the hazard can be removed completely. If that is not reasonably possible, it should consider replacing the hazard with something safer, then look at physical or technical controls, then administrative measures such as procedures or training, and only after that rely on PPE. This order matters because the upper levels generally control the hazard more reliably than the lower levels, which depend more on consistent human behavior. ISO's small-organization guide makes the same point in plain language: each step lower in the hierarchy is less effective than the one above it, although several controls used together can still reduce risk well.

For small enterprises, the hierarchy of controls is useful because it prevents safety decisions from becoming too dependent on reminders, signs, or PPE alone. A business may be tempted to rely on quick fixes, especially when time and resources are limited, but ISO's guidance encourages a more disciplined approach. If a hazard can be designed out, isolated, guarded, ventilated, or otherwise reduced before it reaches the worker, that is usually stronger than relying only on instructions or personal equipment. The result is a more stable and less fragile control system in daily operations.

A practical way to apply the hierarchy is to ask a short sequence of control questions before approving a task or process:

- Can the hazard be eliminated completely?
- Can it be substituted with a safer material, tool, or method?
- Can engineering controls reduce exposure?
- What administrative controls are still needed?
- Is PPE required as an additional protective layer?

Managing Contractors, Purchasing, and Outsourced Activities

ISO 45001 gives clear importance to procurement, contractors, and outsourcing because OH&S risks do not stop at the boundary of directly employed staff. ISO's launch coverage states that the standard places strong emphasis on health and safety management in supply chains so that contracting, procurement, and outsourcing are managed more responsibly. It also notes that outsourced processes and subcontractors feature prominently in the standard, helping organizations identify, control, and continually improve opportunities to reduce or eliminate OH&S risk in the supply chain.

Managing contractors begins with understanding that a contractor is an external organization providing services to the organization in accordance with agreed specifications, terms, and conditions. ISO's FAQ also explains that when a function or process is performed by an external organization on behalf of the organization, it has been outsourced, and that external organization is therefore also a contractor. This distinction matters because not every contractor activity is an outsourced process, but outsourced functions and processes still fall within ISO 45001 control expectations.

For outsourced activities, ISO's FAQ states that outsourced functions and processes must be controlled and that the type and degree of control must be defined within the OH&S management system. ISO also warns that OH&S risks to the organization's own Workers can still arise from related activities connected to outsourced work, such as packaging, loading, and transportation. This means a small enterprise cannot assume that risk disappears simply because part of the work is performed elsewhere. If the outsourced activity affects the organization's work, Workers, or operations, OH&S control still matters.

Purchasing also has an OH&S dimension. ISO's small-organization guide notes that businesses may need to think about how they buy products and services and how that can affect the health and safety of both their own Workers and the people providing those services when they are in the workplace. In practice, this means purchasing decisions should not be based only on cost, speed, or convenience. Bought-in machinery, chemicals, tools, services, maintenance support, temporary labor, and contractor activities can all change workplace risk. A poor purchasing decision can introduce hazards that are difficult to control later.

For a small enterprise, practical control of contractors, purchasing, and outsourced work usually includes three things. First, the organization sets clear safety expectations before work begins. Second, it checks whether the contractor or supplier can work in a way that keeps risks controlled. Third, it monitors whether the agreed controls are actually being followed during delivery, installation, servicing, transport, or on-site work. ISO's small-business guidance also notes that different businesses or contractors may manage health and safety differently, so the key is not forcing identical systems but making sure risks are effectively managed.

Providing and Managing Personal Protective Equipment

Personal protective equipment is an important control measure, but ISO's current safety guidance places it at the bottom of the hierarchy of controls. That does not mean PPE is unimportant. It means PPE is generally less effective than eliminating the hazard, substituting it, or reducing exposure through engineering or administrative means. ISO's guidance is explicit that PPE should be used when other controls are not enough, not as the automatic first solution.

Providing PPE means making suitable protective equipment available for the hazards that remain after higher-level controls have been considered. Depending on the work, this may include helmets, gloves, eye protection, hearing protection, masks, safety footwear, or other protective items. ISO's practical examples include safety shoes and hearing protection as PPE examples, while its current workplace safety guide refers to gear such as helmets, masks, and gloves. The important principle is that PPE should match the actual hazard and working conditions rather than be issued as a generic item with no clear control purpose.

Managing PPE is more than handing equipment to Workers. ISO's current PPE guidance stresses that guidelines should be clear and accessible, regular training and hands-on practice are important for proper use, and worn or damaged PPE should be replaced promptly. ISO also notes more broadly that PPE works best when it is worn correctly, maintained regularly, and supported by clear policies and a workforce that understands why it matters. In practical terms, this means PPE management includes selection, availability, condition, replacement, user understanding, and day-to-day supervision.

For small enterprises, PPE often becomes the most visible sign of OH&S control, but it should not become the only sign. A workplace is not well controlled just because people are wearing protective gear. If guards are missing, ventilation is poor, unsafe methods are accepted, or contractor risks are unmanaged, PPE alone will not provide enough protection. ISO's hierarchy guidance points toward a layered approach: use stronger controls first, then support them with administrative measures and PPE where needed. That is the most reliable way to reduce exposure and improve daily safety performance.

Topic	Main ISO 45001 Direction	Practical Meaning for a Small Enterprise
Hierarchy of controls	Use controls in order of effectiveness, from elimination down to PPE	Do not rely on PPE or procedures first when stronger controls are possible
Contractors and outsourced activities	Control outsourced functions and processes, and define the type and degree of control	Set expectations before work starts and manage risk during contractor or outsourced activity
Purchasing	Consider how purchased products and services affect health and safety	Buy equipment, services, and materials with OH&S impact in mind, not cost alone

PPE	Use as the lowest control level, supported by training and maintenance	Provide suitable PPE, keep it usable, and make sure Workers know how and why to use it

Applying the hierarchy of controls means choosing safety measures in the proper order, starting with the most effective solutions. The first priority is to eliminate the hazard if possible. If that cannot be done, the organization should consider substitution, engineering controls, and administrative controls before relying on personal protective equipment. This approach helps small enterprises reduce risk more effectively and avoid depending only on instructions or PPE.

Managing contractors, purchasing, and outsourced activities is essential because OH&S risks can also enter the workplace through external parties, bought-in products, or outsourced work. Small enterprises should set clear safety expectations, review risks before work begins, and make sure outside activities do not weaken workplace control. Purchasing decisions should also consider safety impacts, not just cost or convenience.

Providing and managing personal protective equipment is important when other controls cannot fully remove the risk. PPE should be suitable for the specific hazard, available when needed, properly used, maintained in good condition, and replaced when damaged. It should always support stronger controls, not replace them.

A Personal Note from the Author

In this chapter, I wanted to show that OH&S becomes meaningful only when it is translated into everyday practice. Policies and plans are important, but they do not protect people on their own. Real protection comes when the organization provides the right resources, builds competence, strengthens communication, controls documentation, develops safe work practices, manages external activities properly, and applies controls in a practical and disciplined way.

I also wanted to highlight that small enterprises do not need complicated systems to manage OH&S effectively. What they need is clarity, consistency, and follow-through. When safety requirements are supported by proper resources, clear communication, suitable controls, and responsible supervision, they become part of normal work rather than an additional burden. My intention in this chapter is to help readers see that daily practice is where OH&S succeeds or fails.

CHAPTER 14

Preparing for Emergencies and Checking Performance

Chapter Four brings together two ideas that are closely connected in ISO 45001:2018: being ready for abnormal or dangerous situations, and checking whether the OH&S system is actually working. ISO describes emergency planning as one of the key elements of the standard and also includes auditing and review as part of the overall OH&S management system framework. That structure matters because emergency preparedness is not only about reacting when something goes wrong. It is also about planning ahead, testing readiness, learning from results, and improving the system over time through the Plan-Do-Check-Act approach.

Identifying Possible Emergency Situations

Identifying possible emergency situations means determining which events could suddenly threaten people, disrupt work, or damage the organization's ability to operate safely. ISO's current OH&S guidance stresses that emergencies do not wait for the right moment and can include events such as fire, flood, cyberattack, or a health crisis. ISO 45001 requires organizations to treat emergency planning as part of a structured OH&S management system rather than as an afterthought. For a small enterprise, this means the business should think beyond ordinary daily risks and consider what kinds of urgent, abnormal events could realistically arise from its activities, equipment, materials, premises, utilities, contractors, and external environment.

This identification step is important because emergency preparedness can only be effective when it is based on real exposure. A workplace with flammable materials, forklifts, compressed gas, electrical panels, working at height, lone work, visitors, or contractor activity may face very different emergency scenarios from a small office or retail outlet. ISO's broader OH&S approach is built on managing risks through understanding hazards and planning controls, and that same logic applies to emergencies: the organization should identify what could happen, who could be affected, and what response capability would be needed if the event occurred.

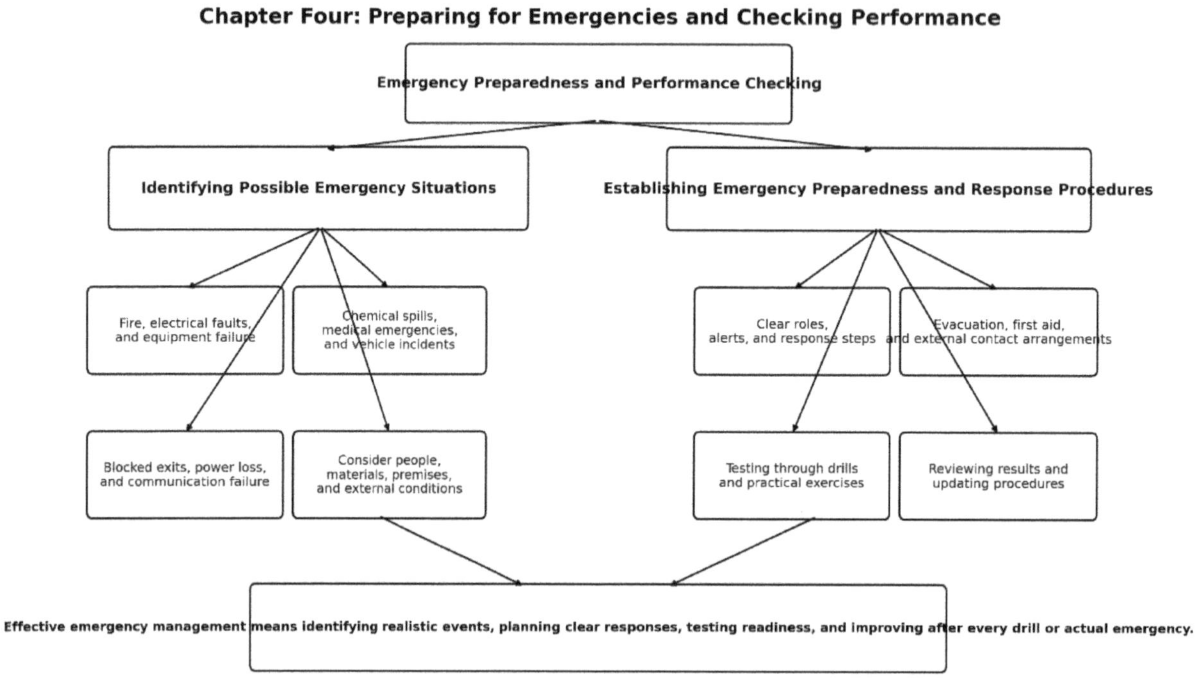

In practical terms, identifying emergencies is not limited to dramatic disasters. It also includes events that may be smaller in scale but still serious for a small business, such as loss of power affecting safe shutdown, chemical spills, vehicle incidents, medical emergencies, equipment failure, blocked exits, or communication breakdown during an evacuation. ISO's current guidance describes preparedness as more than evacuation routes and drills and points to the need for backup systems, trained response teams, and crisis communication protocols. That broader view is especially useful for small enterprises because even a relatively limited event can cause major disruption when staffing and resources are tight.

Establishing Emergency Preparedness and Response Procedures

Once possible emergencies have been identified, the organization needs response procedures that are clear, practical, and usable under pressure. ISO 45001 includes a specific requirement for emergency preparedness and response, and the ISO/UNIDO Q&A explains that organizations are required to test their plans periodically and evaluate their effectiveness through those tests and through any actual emergencies. That means emergency procedures are not simply written instructions to satisfy documentation needs. They are working arrangements that must be capable of guiding a real response.

An effective emergency procedure usually answers the essential operational questions: what must be done, who must do it, how people will be alerted, how the situation will be controlled, how evacuation or sheltering will take place if needed, how external support will be contacted, and how the business will account for people and restore control afterward. ISO's current OH&S guidance describes preparedness as a living, evolving mindset and highlights readiness measures such as trained response teams, backup systems, and crisis communication protocols. In a small enterprise, these procedures do not need to be complex, but they do need to be specific enough that people can act quickly and consistently.

A key point in ISO guidance is that emergency procedures must be tested and then evaluated. The ISO/UNIDO Q&A states that plans should be tested periodically and that their effectiveness should be evaluated based on drills and any real emergencies. It also notes that timeliness is an important attribute when judging whether a plan is effective. For small enterprises, this is critical. A procedure may look adequate on paper but still fail if alarms are unclear, exits are blocked, roles are misunderstood, first-aid support is delayed, or coordination with contractors and visitors breaks down during the event.

Emergency procedures should also be revised when the organization learns something new. ISO's broader management system model is based on continual improvement, so emergency preparedness is expected to improve through testing, review, and corrective action rather than remain static. If a drill reveals weak communication, poor evacuation discipline, missing equipment, or confusion about responsibilities, those Observations and recommendations must be documented; implementation is per organization discretion. should lead to updates in the procedure, additional training, or better resource allocation. This is one of the clearest ways that preparing for emergencies connects directly with checking performance.

For small enterprises, the strongest emergency arrangements are usually the ones that are realistic, role-based, and regularly practiced. They should reflect the actual workplace layout, staffing pattern, shift arrangements, contractor presence, and available resources. ISO 45001 applies to organizations of all sizes, so the standard does not require one uniform emergency model for every business; it requires an effective one. A small enterprise therefore benefits most from emergency procedures that are simple enough to use in real time, but disciplined enough to protect Workers, coordinate response, and support improvement after the event.

Testing, Reviewing, and Updating Emergency Arrangements

Under ISO 45001:2018, emergency preparedness is not complete when a procedure is written. The organization is expected to **test its emergency plans periodically** and **evaluate their effectiveness** using the results of those tests and any real emergencies. ISO's official Q&A makes this point directly and adds that **timeliness** is an important part of judging whether an emergency plan is effective. This means emergency arrangements must be treated as living controls, not static documents.

Testing emergency arrangements helps the organization confirm whether people can actually respond as intended under pressure. In a small enterprise, this may reveal very practical weaknesses, such as unclear alarms, slow communication, blocked escape routes, poor role understanding, missing equipment, or weak coordination with contractors and visitors. ISO also highlights that emergency preparedness and response protocols strengthen organizational resilience, which shows that testing is part of maintaining real readiness rather than only meeting a documentation requirement.

Reviewing emergency arrangements means looking carefully at what the drill, exercise, or real event has shown. The organization should consider whether the response was timely, whether responsibilities were understood, whether communication worked, whether people knew where to go, and whether the available resources were sufficient. Review is important because a plan may appear adequate on paper but still fail in actual use. The value of testing is therefore not only in carrying out the exercise, but in learning from it in a disciplined way.

Updating emergency arrangements is the improvement step that follows review. If testing or a real event shows confusion, delay, missing controls, or unrealistic assumptions, the organization should revise the procedure, provide additional training, improve communication methods, adjust responsibilities, or strengthen resources. This fits the ISO 45001 Plan-Do-Check-Act logic, where emergency preparedness is strengthened through repeated testing, evaluation, and corrective follow-up rather than left unchanged after the first version is written.

A practical emergency review usually checks:

- whether the response was fast enough
- whether people understood their roles
- whether communication worked clearly
- whether equipment and resources were available
- whether the procedure needs revision after the test or event

Monitoring, Measuring, and Evaluating OH&S Performance

ISO 45001 places this topic in **Clause 9.1: Monitoring, measurement, analysis and performance evaluation**. ISO's browsing platform shows that performance evaluation is a formal part of the standard, and ISO's current OH&S guidance explains that organizations should manage OH&S performance through **ongoing monitoring, measurement and evaluation**, along with regular high-level reviews of the effectiveness of controls. This means the organization must not assume its OH&S system is working simply because procedures exist. It has to check performance in a structured way.

Monitoring means observing whether OH&S activities and controls are functioning as intended during normal work. Measuring means collecting information that shows performance or results. Evaluation means interpreting that information to decide whether the OH&S system is effective, whether risks are controlled, and whether action is needed. ISO's harmonized management-system guidance states that organizations should determine what needs to be monitored and measured, the methods to use, when monitoring and measuring will occur, and when the results will be analyzed and evaluated. Although that guidance is generic across ISO management system standards, it reflects the same Clause 9 logic used in ISO 45001.

For a small enterprise, monitoring and performance evaluation should stay practical. It may include workplace inspections, supervisor observations, incident and near-miss trends, overdue corrective actions, equipment condition, training completion, emergency drill results, and recurring unsafe conditions. The goal is not to collect large amounts of data for its own sake, but to understand whether OH&S controls are working in real operations and whether risks are being kept under control. ISO's hazards guidance specifically links ongoing monitoring and evaluation with the need to review the effectiveness of controls and take corrective action where necessary.

Performance evaluation is also important because it supports decision-making. When management knows where controls are weak, where actions are overdue, where incidents are recurring, or where emergency readiness is poor, it can direct resources and corrective action more effectively. ISO's management-system guidance explains that information gained through monitoring, measurement, analysis and evaluation is intended to support decision-making and drive continual improvement. In OH&S terms, this helps the organization move from reactive correction to better ongoing control.

A practical OH&S performance check may include:

- inspection Observations and recommendations must be documented; implementation is per organization discretion.
- incident and near-miss patterns
- completion of corrective actions
- effectiveness of emergency drills
- recurring weaknesses in daily controls

Area	What the Organization Checks	Practical Purpose
Emergency testing	Whether drills and response exercises work in real conditions	Confirms readiness and exposes weak points
Emergency review	Whether the response was timely, clear, and effective	Identifies what failed and what must improve
Emergency updating	Whether procedures, roles, training, or resources need revision	Keeps preparedness current and usable
OH&S monitoring	Whether controls are being applied during normal work	Shows if daily safety arrangements are functioning
OH&S measurement and evaluation	Whether results indicate effective control or the need for action	Supports decisions, corrective action, and continual improvement

This table shows how emergency follow-up and OH&S performance evaluation work together: one checks readiness for abnormal events, and the other checks how well the OH&S system is functioning in everyday operations. Both are essential to the "Check" and "Act" parts of ISO 45001.

Evaluating Compliance with Requirements

In ISO 45001:2018, compliance is not treated as a passive assumption. ISO identifies **legal and regulatory compliance** as a key element of the OH&S management system, and its current explanation of the standard includes the **determination of legal and other requirements** and **monitoring, measurement and performance evaluation** among the core requirements. The intended outcomes of the OH&S management system also include the **fulfilment of legal requirements and other requirements**. Taken together, this means the organization is expected not only to identify applicable requirements, but also to check whether those requirements are actually being met in practice.

Evaluating compliance with requirements means reviewing whether the organization's OH&S activities, controls, records, and working practices match the legal and other requirements it has determined to be applicable. In practical terms, this is not limited to asking whether a policy exists. It involves checking whether legal duties and other accepted requirements have been translated into operational control, communication, supervision, training, maintenance, emergency readiness, reporting, and follow-up. This is a reasonable interpretation of ISO 45001's intended outcomes and its performance-evaluation logic.

For a small enterprise, compliance evaluation should be practical and repeatable. It may involve reviewing whether inspections are being completed, whether required training has taken place, whether emergency equipment is available, whether contractor controls are followed, whether statutory checks are current, and whether corrective actions linked to compliance issues have been closed. The purpose is to confirm that compliance is active in daily operations rather than existing only in documents or management intentions.

A practical compliance evaluation often checks:

- whether all relevant legal and other requirements have been identified

- whether those requirements have been built into procedures and controls
- whether actual workplace practice matches those requirements
- whether gaps, delays, or repeated weaknesses are being corrected

Conducting Internal Audits

Internal audit is one of the main ways an organization checks whether its OH&S management system is functioning as intended. ISO 45001 explicitly includes **auditing and review** within the framework of the standard, and ISO's harmonized management-system structure states that organizations shall conduct internal audits at planned intervals to provide information on whether the management system conforms to the organization's own requirements and to the requirements of the standard, and whether it is effectively implemented and maintained.

This makes internal audit more than a document check. It is a structured process for examining whether OH&S arrangements are working across real activities, responsibilities, and controls. An effective internal audit looks at how the system operates in practice: whether requirements are understood, whether controls are applied, whether records reflect reality, whether previous issues were resolved, and whether the OH&S system is delivering the results expected from it. ISO's harmonized structure also requires the organization to plan, establish, implement, and maintain an audit programme, including frequency, methods, responsibilities, planning requirements, and reporting.

The audit programme should be planned intelligently rather than applied in the same way everywhere. ISO requires the organization to consider the importance of the processes concerned and the results of previous audits when establishing the programme. This means higher-risk areas, weak controls, recurring nonconformities, and problem-prone operations should receive more attention. In a small enterprise, that helps management use limited time and resources where audit activity is most useful.

Objectivity is also essential. ISO requires the organization to define the audit objectives, criteria, and scope for each audit, and to select auditors and conduct audits in a way that ensures objectivity and the impartiality of the audit process. In small businesses, where people may hold several roles, this principle is especially important. The audit should not become a routine self-approval exercise. It should provide reliable information that management can use to improve control. ISO's harmonized structure also notes ISO 19011 as the guidance standard for auditing management systems.

A sound internal audit normally addresses:

- what will be audited and why
- which requirements and processes are being checked
- who will audit and how objectivity will be protected
- what Observations and recommendations must be documented; implementation is per organization discretion., conclusions, and follow-up actions will be reported

Carrying Out Management Review

Management review is the formal point at which top management steps back and evaluates whether the OH&S management system remains suitable, adequate, and effective. ISO's harmonized management-system structure states that top management shall review the management system at planned intervals for this purpose, and the guidance explains that this review is the mechanism by which top management drives changes to the system and directs continual-improvement priorities. ISO 45001 also places strong emphasis on leadership commitment and on review as a core part of the OH&S framework.

This means management review is not simply a meeting to receive information. It is a leadership activity in which management examines whether the OH&S system still fits the organization's conditions, whether its controls are effective, whether key problems are being addressed, and whether changes are needed. In small enterprises, this review is particularly valuable because business conditions can shift quickly through growth, staffing changes, new work, new equipment, changing contractors, or operational pressure. Management review helps prevent the system from becoming outdated or disconnected from actual business conditions.

ISO's harmonized structure identifies the main review inputs. These include the status of actions from previous management reviews, changes in internal and external issues, changes in the needs and expectations of relevant interested parties, and information on performance trends such as nonconformities and corrective actions, monitoring and measurement results, and audit results. For an OH&S management system, these inputs allow top management to see not only isolated problems but also patterns that may show weakness, drift, or opportunities for improvement.

The review must also lead to outputs. ISO requires that the results of management review include decisions related to continual-improvement opportunities and any need for changes to the management system, with documented information kept as evidence of the review results. In practical terms, this means the review should end with direction: what will change, what will be prioritized, what resources are needed, what risks require stronger control, and how the organization will move forward. Without those outcomes, management review becomes discussion rather than management action.

Topic	Main Purpose	Practical Result
Evaluating compliance with requirements	Check whether legal and other requirements are actually being fulfilled	Confirms that compliance exists in practice, not only on paper
Conducting internal audits	Determine whether the OH&S management system conforms and is effectively implemented and maintained	Provides reliable Observations and recommendations must be documented; implementation is per organization discretion. for correction, control, and improvement
Carrying out management review	Enable top management to review continuing suitability, adequacy, and effectiveness	Produces decisions on improvement priorities and needed system changes

These three activities work together in the **Check** and **Act** parts of ISO 45001. Compliance evaluation confirms whether obligations are being met, internal audit checks whether the system is functioning properly, and management review ensures that top management uses this information to direct improvement and keep the OH&S management system effective over time.

A Personal Note from the Author

In this chapter, I wanted to show that emergency preparedness and performance evaluation are not separate management activities. They are both essential to knowing whether an OH&S system is truly working. A small enterprise may have emergency procedures, inspection records, or audit Observations and recommendations must be documented; implementation is per organization discretion., but these only become meaningful when they are tested, reviewed carefully, and used to guide better decisions. My intention was to emphasize that real safety management depends on readiness, honest evaluation, and consistent follow-through.

I also wanted to make clear that compliance, internal audits, and management review should not be treated as routine formalities. They are practical tools for checking whether responsibilities are being carried out, whether controls are effective, and whether leadership is responding properly to risk and change. In small enterprises, where resources are limited and the effects of mistakes can be serious, these activities help management stay informed, maintain control, and strengthen the OH&S system over time.

CHAPTER 15

Driving Improvement and Creating a Proactive Safety Culture

Chapter Five focuses on the improvement side of ISO 45001:2018. ISO identifies **incident investigation** and **continual improvement** as key elements of the standard, and its current explanation of ISO 45001 says organizations use the system to monitor incidents and near misses, review results, and improve regularly. This means the OH&S management system is not meant to stop once hazards are identified and controls are introduced. It is supposed to keep learning from what goes wrong, what almost goes wrong, and what shows early signs of weakness in daily work.

A proactive safety culture is closely connected to this improvement process. ISO's practical guide for small organizations explains that implementing ISO 45001 can support a **more cooperative culture** with stronger worker engagement and empowerment, better management of OH&S risks, and a better approach to incident investigation. In small enterprises, this matters greatly because safety culture is often shaped directly by the visible actions of owners, managers, and supervisors. When the business encourages reporting, responds fairly to concerns, investigates problems properly, and follows through with real action, Workers are more likely to speak up early and participate honestly in safety management.

Reporting Hazards, Incidents, and Unsafe Conditions

Reporting is one of the strongest practical tools in a functioning OH&S management system because it gives the organization early information about where risk exists and where controls may be failing. ISO's current guidance on ISO 45001 lists identifying hazards and unsafe conditions, monitoring incidents and near misses, and planning corrective and Preventive Emergency arrangements should be communicated clearly to Workers among the practical first steps for an effective system. This makes reporting more than an administrative requirement. It is one of the main ways the organization keeps the system connected to real workplace conditions.

A **hazard** is any work-related situation, activity, or process that can harm people. ISO's occupational safety guidance gives a broad understanding of hazards, including obvious physical dangers as well as workload and resource-related issues. An **incident** includes events such as accidents that cause injury and also near misses that did not result in harm this time but still reveal exposure to risk. **Unsafe conditions** are the workplace situations that increase the chance of harm, such as damaged equipment, blocked exits, missing guards, poor housekeeping, weak supervision, unclear communication, or uncontrolled contractor activity. Reporting these conditions early is important because hazards that are not eliminated or otherwise controlled create OH&S risks.

For small enterprises, good reporting helps shift safety management from reaction to prevention. ISO's current explanation of the standard says organizations use ISO 45001 to move from reactive safety measures to proactive risk management. In practical terms, this means the organization should not wait until a serious injury, enforcement issue, or operational disruption occurs before it pays attention. If Workers can report hazards, unsafe conditions, and near misses promptly, the business gets an earlier chance to investigate, correct, and improve. That is one of the clearest signs of a proactive safety culture.

Reporting also supports worker participation, which ISO treats as a core requirement of the standard. ISO's official Q&A makes clear that participation and consultation of Workers is required under ISO 45001, and ISO's public material highlights worker involvement in safety discussions as a practical first step. A reporting system therefore should not make Workers feel that they are complaining, creating trouble, or inviting blame. It should make them feel that they are contributing useful information to help protect people and improve work. In small enterprises, where daily relationships are close and management behavior is highly visible, this point is especially important.

A practical reporting process in a small enterprise usually works best when it is simple, quick, and trusted. Workers should know what to report, who to report to, and what will happen next. Management should respond consistently, record what matters, and follow up visibly enough that Workers can see reporting leads to action.

When reports disappear into silence, people quickly stop raising concerns. When reports lead to fair review and real correction, reporting becomes part of normal OH&S practice rather than an exceptional event. This aligns with ISO's broader emphasis on leadership, worker participation, performance evaluation, and continual improvement.

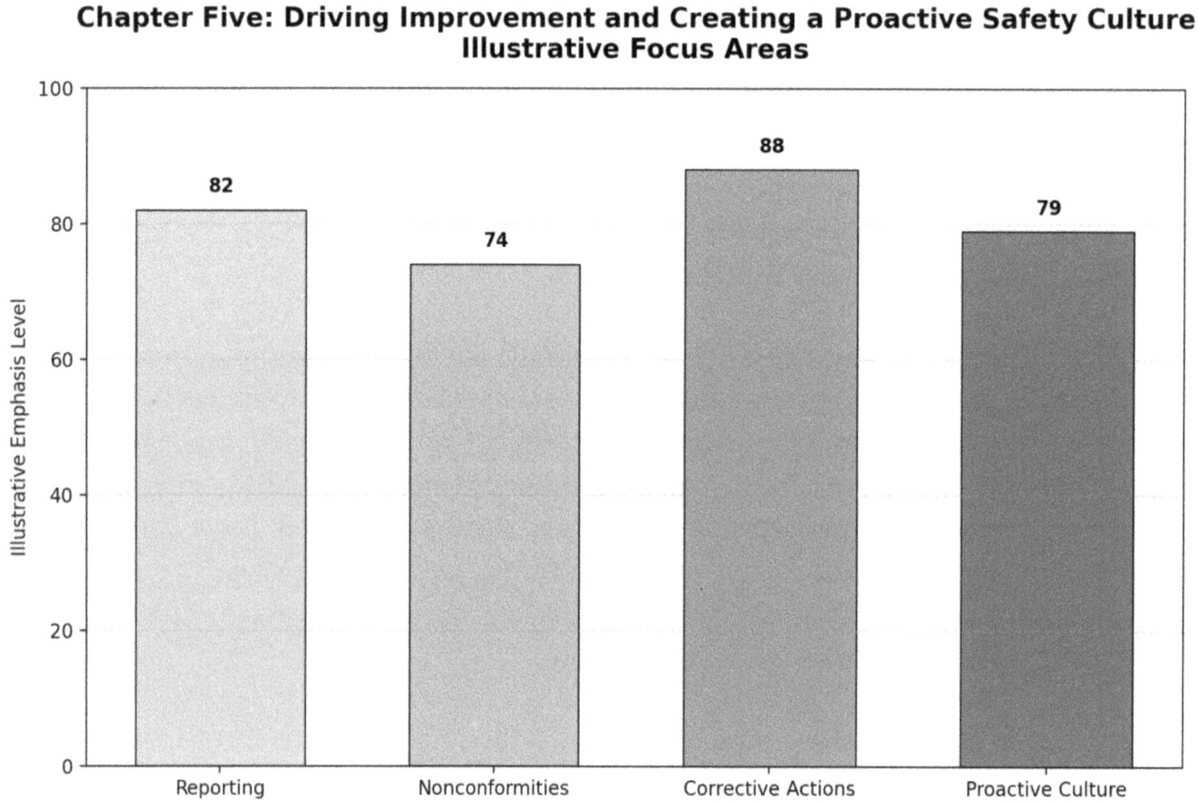

Managing Nonconformities and Corrective Actions

Nonconformity and corrective action sit in **Clause 10: Improvement** in ISO 45001. ISO's public explanation of the standard identifies **continual improvement** as a core requirement, and the practical small-organization guide lists **nonconformity and corrective action** among the management-system elements organizations need to handle when implementing ISO 45001. This means that when the OH&S system fails to meet a requirement, control, or intended result, the organization is expected to deal with that failure systematically rather than informally.

A **nonconformity** in OH&S terms can take many forms. It may be a missed inspection, an overdue corrective action, failure to follow a safe work practice, poor emergency readiness, weak contractor control, missing training, inaccurate records, or a gap between legal or internal requirements and what is actually happening at the workplace. The key point is that the system is not working as intended in that area. ISO's management-system logic does not treat these failures as isolated annoyances. They are signals that a control, process, responsibility, or resource arrangement needs attention.

Managing a nonconformity begins with recognizing it and responding to it in a controlled way. The immediate response may include correcting the situation, reducing any OH&S risk created by the problem, and preventing further exposure while the issue is reviewed. That first response is important, but it is not enough by itself. If the organization only fixes the visible symptom and does not consider why the problem happened, the same weakness may return in a slightly different form. This is why ISO 45001 connects corrective action with continual improvement rather than treating it as a quick repair.

Corrective action is the action taken to address the cause of a nonconformity so that it does not recur. In practical OH&S management, this means looking beyond the immediate event and asking whether the real cause was weak supervision, poor communication, unclear responsibilities, unrealistic procedures, insufficient competence, poor maintenance, inadequate resources, or failure to learn from earlier warning signs. ISO's current practical guidance explicitly includes planning corrective actions and Emergency arrangements should be communicated clearly to Workers. as part of building an effective OH&S management system. planning corrective actions and Emergency arrangements should be communicated clearly to Workers. as part of building an effective OH&S management system. **with "and and planning corrective actions and Emergency arrangements should be communicated clearly to Workers.. actions and preventive measures. actions and preventive measures.** as part of building an effective OH&S management system. For small enterprises, this cause-based approach is essential because repeating the same failures wastes scarce time and resources and weakens trust in the system.

Good corrective action also depends on follow-through. The organization should assign responsibility, set realistic timeframes, provide any needed resources, and later check whether the action actually solved the problem. If an issue reappears, that is evidence that the action was incomplete, poorly implemented, or directed at the wrong cause. This links corrective action directly to the wider ISO 45001 cycle of monitoring, review, and continual improvement. A corrective action is valuable only when it strengthens control in practice, not when it simply closes a record.

In a proactive safety culture, nonconformities are treated as useful information, not as embarrassment to be hidden. That does not mean failures are accepted casually. It means the organization responds to them honestly and uses them to improve the OH&S system. ISO's small-organization guide connects ISO 45001 with better incident investigation, better identification of opportunities to improve OH&S management, and a more cooperative culture. For small enterprises, this is one of the most practical benefits of managing nonconformity and corrective action well: the system becomes more open, more disciplined, and less likely to repeat preventable failures.

Taken together, reporting hazards, incidents, and unsafe conditions gives the organization the information it needs, while managing nonconformities and corrective actions gives it a structured way to respond and improve. Reporting without corrective action creates frustration. Corrective action without honest reporting leaves problems hidden. ISO 45001 expects both, because improvement depends on seeing weakness clearly and then dealing with it in a disciplined, preventive way.

Learning from Incidents and Audit Results

Learning from incidents and audit results is a central part of improvement in ISO 45001:2018. ISO identifies **incident investigation** and **continual improvement** as key elements of the standard, and it states that the intended outcomes of the OH&S management system include continually improving OH&S performance and fulfilling legal and other requirements. This means the organization is expected to use incidents, near misses, nonconformities, and audit Observations and recommendations must be documented; implementation is per organization discretion. as sources of learning, not just as records to close.

An incident should be treated as information about how the system performed under real conditions. It may show that a hazard was not identified fully, a control was weak, a procedure was unclear, supervision was insufficient, communication failed, or resources were inadequate. In small enterprises, this learning is especially valuable because even one incident can reveal weaknesses that affect several parts of the business at once, including training, maintenance, contractor control, or emergency readiness. ISO's small-organization guidance highlights a better approach to incident investigation and better identification of opportunities to improve OH&S management as important outcomes of using ISO 45001 well.

Audit results provide a different but equally important source of learning. Incidents show what happened during actual work, while audits show whether the management system is being followed and whether controls are in place before a serious event occurs. Audit Observations and recommendations must be documented; implementation is per organization discretion. can reveal missing documentation, overdue actions, poor implementation of procedures, inconsistent supervision, weak performance monitoring, or gaps between planned arrangements and daily practice. When used properly, audit results help the organization detect system weakness before it causes harm.

The real value comes from combining both sources. If incident records, near-miss reports, inspection Observations and recommendations must be documented; implementation is per organization discretion., and audit results point to the same weakness, the organization has strong evidence that the issue is systemic rather than accidental. For example, repeated manual-handling incidents together with audit Observations and recommendations must be documented; implementation is per organization discretion. about weak training and poor supervision would show that improvement is needed at both task level and system level. ISO 45001's Plan-Do-Check-Act structure supports exactly this kind of learning cycle: review what happened, identify what failed, correct it, and improve the system so the same weakness is less likely to return.

A sound learning process should not focus only on blame. It should ask what the organization needs to understand about the event or finding and what changes are required in order to strengthen control. In practical terms, that may involve reviewing responsibilities, competence, communication, safe work practices, contractor arrangements, maintenance, emergency preparation, or management follow-up. ISO's small-business guidance links ISO 45001 with a more cooperative culture and stronger worker engagement, which supports a learning approach that is fair, disciplined, and improvement-focused.

A practical review after an incident or audit usually asks:

- What happened, or what weakness was found?
- Which control failed, was missing, or was not followed?
- Was this an isolated issue or part of a wider pattern?
- What action is needed to stop recurrence and strengthen the system?

Source of Learning	What It Can Reveal	Why It Matters
Incidents and near misses	Failures in real working conditions	Shows where people, tasks, or controls were exposed to actual risk
Audit Observations and recommendations must be documented; implementation is per organization discretion.	Gaps in implementation, records, or system control	Helps detect weakness before it leads to harm
Repeated patterns across both	Systemic weakness rather than isolated error	Supports stronger corrective action and better prioritization
Worker feedback and reporting	Early warning signs and practical obstacles	Keeps improvement connected to real work

This table shows that incidents and audit results should not be treated separately. Together, they give management a clearer picture of where the OH&S management system is strong, where it is weak, and where improvement should be directed first.

Continually Improving the OH&S Management System

Continual improvement means the OH&S management system does not remain fixed after it has been established. ISO states that the framework is designed so the system can continually improve and evolve, and the intended outcomes of the system include continual improvement of OH&S performance. This means the organization should keep strengthening the suitability, adequacy, and effectiveness of its OH&S arrangements as work, risks, and business conditions change.

In practical terms, continual improvement is not a single project or a once-a-year review. It is an ongoing management process driven by evidence. That evidence may come from incidents, audits, inspections, worker reports, performance trends, legal-compliance checks, emergency tests, and management review. Each of these gives the organization information about what is working, what is weak, and what should be improved next. ISO's public guidance explains that organizations use ISO 45001 to monitor incidents and near misses, review results, and improve regularly, which reflects this continuous cycle of learning and strengthening.

For small enterprises, continual improvement should remain practical and proportionate. It does not require constant redesign of the whole system. More often, it means improving the areas that matter most: making a procedure clearer, strengthening supervision, updating training, improving contractor control, replacing weak equipment, closing overdue actions faster, or improving communication during change. ISO's small-organization guidance notes benefits such as better management of OH&S risks, better identification of opportunities to improve OH&S management, and a more cooperative culture with greater worker engagement and empowerment.

A strong improvement process also depends on follow-through. If the organization identifies lessons but does not assign responsibility, provide resources, set timeframes, and verify effectiveness, improvement remains incomplete. Continual improvement therefore depends on both action and verification. The system becomes stronger only when the organization can see that the change actually reduced risk, corrected the weakness, or improved OH&S performance in practice. That is consistent with ISO 45001's performance-oriented approach and its use of the Plan-Do-Check-Act model.

Continual improvement also supports culture. When Workers see that reports lead to action, audit Observations and recommendations must be documented; implementation is per organization discretion. lead to better controls, and management takes follow-up seriously, confidence in the OH&S system increases. Over time, this helps create a more proactive culture in which people are more willing to report concerns, participate in solutions, and maintain safer work practices. ISO's small-business guidance specifically links improvement with better worker engagement and a more cooperative culture, which is particularly important in small enterprises where management behavior is highly visible.

A practical continual-improvement process usually includes:

- identifying lessons from incidents, audits, and routine monitoring
- deciding which actions are most important and most urgent
- assigning responsibility and resources for implementation
- checking whether the action actually improved control or performance

Improvement Area	Typical Input	Improvement Result
Procedures and controls	Incident Observations and recommendations must be documented; implementation is per organization discretion., near misses, unsafe-condition reports	Safer and clearer day-to-day work methods
Competence and awareness	Audit Observations and recommendations must be documented; implementation is per organization discretion., supervision results, recurring mistakes	Better understanding and more reliable task performance
Resources and equipment	Repeated failures, maintenance issues, emergency tests	Stronger operational control and reduced exposure
Management follow-up	Trends, overdue actions, review results	More disciplined correction and better long-term OH&S performance

Taken together, learning from incidents and audit results gives the organization the evidence it needs, while continual improvement turns that evidence into stronger control, better performance, and a more proactive safety culture. In ISO 45001, improvement is not optional extra work at the end of the system. It is one of the system's intended outcomes.

Creating a Positive Reporting Culture

A positive reporting culture is an OH&S environment in which Workers feel able and expected to report hazards, near misses, incidents, unsafe conditions, and OH&S concerns in time for the organization to act. In ISO 45001:2018, this idea is closely connected to **worker consultation and participation**, **incident investigation**, **performance evaluation**, and **continual improvement**. ISO's public guidance explains that organizations using ISO 45001 should identify hazards and unsafe conditions, monitor incidents and near misses, involve Workers in safety discussions, and improve regularly. That means reporting is not a side activity; it is one of the main ways the system stays connected to what is really happening at work.

A reporting culture becomes positive when people believe that reporting has value, that management will listen, and that concerns will lead to fair review and practical follow-up. ISO's small-organization guidance links ISO 45001 with a **more cooperative culture**, stronger worker engagement and empowerment, and a better approach to incident investigation. That matters for small enterprises because Workers are often closest to changing conditions, routine shortcuts, equipment issues, workload pressure, and emerging risks. If they stay silent, the organization loses some of its earliest warning signs.

Leadership plays a decisive role in shaping this culture. ISO's 2018 launch material says ISO 45001 places more focus on **leadership** and **worker participation**, with top management expected to take a visible, directing role and integrate the OH&S system into overall business processes. In practical terms, Workers watch how managers react when a problem is raised. If reports are ignored, delayed, or treated defensively, reporting will decline. If leaders respond seriously, ask useful questions, remove obstacles, and follow through, reporting becomes part of normal business discipline rather than an uncomfortable exception.

A positive reporting culture also depends on trust and clarity. Workers should know what kinds of issues to report, how to report them, who receives the information, and what happens next. ISO's official Q&A makes clear that worker consultation and participation is required in ISO 45001, and ISO's current guidance highlights the importance of effective communication and worker involvement in managing OH&S risks. From a practical management perspective, this means the reporting process should be simple enough to use quickly and clear enough that nobody is left unsure whether a concern is worth raising.

Another essential feature is visible follow-up. Reporting culture weakens when Workers repeatedly raise issues but see no action, no feedback, and no learning. It becomes stronger when reports are reviewed, causes are considered, corrective action is assigned, and improvements are communicated back to the workforce. ISO's management-system model is built around self-evaluation, correction, and continual improvement, and ISO 45001 specifically connects incident investigation, nonconformity, corrective action, and continual improvement. A strong reporting culture therefore does not end when a report is submitted; it continues until the organization has learned from the information and used it to strengthen control.

In small enterprises, a positive reporting culture is often one of the clearest signs that OH&S is being managed proactively rather than reactively. It helps the business detect risk earlier, correct weak controls before harm occurs, improve communication between management and Workers, and support a more cooperative and accountable safety culture. When people report openly and management responds responsibly, the OH&S management system becomes more practical, more credible, and more effective over time.

A positive reporting culture in practice usually has four visible features. People know what to report. They can report it without unnecessary difficulty. Management responds in a timely and fair way. The organization uses the information to improve controls, not just to create records. Those four features turn reporting into a working part of ISO 45001 rather than a formal exercise.

Encouraging Safety Discussions and Worker Feedback

Encouraging safety discussions and worker feedback is a practical expression of ISO 45001:2018's requirement for **consultation and participation of Workers**. ISO's official material states that worker participation is a key element of the standard, and its public guidance specifically lists **involving Workers in safety discussions** as one of the practical first steps for strengthening occupational health and safety management.

In practical terms, safety discussions are the regular opportunities Workers have to speak about hazards, unsafe conditions, changing work methods, near misses, controls, and suggestions for improvement. Worker feedback is the information they provide from their direct experience of the job. This matters because Workers are often the first to notice weak controls, unrealistic procedures, unclear instructions, workload pressure, or equipment problems. ISO's 2018 launch coverage explains that ISO 45001 places greater emphasis on Workers actively participating in the development, planning, implementation, and continual improvement of the OH&S management system.

For small enterprises, these discussions are especially valuable because communication lines are shorter and management is usually closer to daily operations. A small business may not need complex consultation structures, but it still needs a dependable way for Workers to raise concerns and contribute ideas. ISO's official Q&A states that participation and consultation of Workers is a requirement of ISO 45001, and top management must ensure this takes place if the organization wishes to claim conformity with the standard.

Safety discussions should not be limited to incidents after they happen. They are most useful when they take place before work starts, during change, after inspections, during toolbox talks, after near misses, and whenever Workers see something that could lead to harm. ISO's recent occupational safety guidance notes that ISO 45001 helps embed a culture of continuous improvement and empowers Workers to play an active role in creating safer workplaces.

A strong feedback process also depends on management behavior. ISO's launch coverage says top management should promote a positive culture, communicate what needs to be done and why it matters, and take an active role in the OH&S system. When leaders listen, respond respectfully, and act on valid concerns, Workers are more likely to speak openly. When feedback is ignored or treated defensively, participation usually weakens.

A practical safety discussion process usually includes:

- regular short discussions on hazards, changes, and controls
- clear ways for Workers to raise concerns and suggestions
- visible management response and follow-up
- feedback to Workers on what action was taken and why

Area	What It Means	Why It Matters in a Small Enterprise
Safety discussions	Workers and supervisors talk about hazards, controls, and work conditions	Helps identify problems early and keep controls realistic
Worker feedback	Workers share observations, concerns, and improvement ideas	Brings real task-level knowledge into OH&S decisions
Consultation	Management seeks worker views before decisions affecting OH&S	Improves trust and strengthens practical control

Participation	Workers actively contribute to implementation and improvement	Makes the OH&S system more accurate and usable in daily work

Encouraging safety discussions and worker feedback also supports a proactive culture. It helps the business move from waiting for incidents to learning from early warning signs. In ISO 45001 terms, this strengthens consultation, participation, communication, performance evaluation, and continual improvement across the OH&S management system.

Promoting a Proactive Approach to OH&S in Small Enterprises

Promoting a proactive approach to occupational health and safety means managing risk before harm occurs, rather than waiting for an incident, complaint, or inspection finding to force action. ISO's current OH&S guidance describes proactive risk management as a practical way to safeguard people, operations, and continuity, and its occupational safety guide explains that a proactive approach involves identifying workplace hazards, understanding risks, and putting tangible measures in place to prevent harm. In ISO 45001:2018, this approach is built into the management system itself through based on based on risk assessment methods defined by the organization in line with ISO 45001:2018 Clause 6.1.2" to align with risk-based thinking., risk assessment, operational control, worker participation, performance evaluation, and continual improvement.

For small enterprises, this approach is especially important because the effects of failure are often more immediate and more severe than in larger organizations. ISO's small-organization guidance explains that the primary focus of an OH&S management system is to prevent work-related injury, illness, or loss of life, and it notes that the requirements of ISO 45001 assist small organizations as much as larger ones. In a small business, a single serious injury, recurring unsafe practice, or weak control can quickly affect staffing, productivity, customer service, morale, and financial stability. A proactive approach helps reduce that vulnerability by encouraging earlier action and better operational discipline.

A proactive approach begins with seeing OH&S as part of normal business management rather than as a separate reaction to problems. ISO states that ISO 45001 provides a framework for systematically assessing hazards and implementing risk-control measures, and its current practical guidance says organizations use the standard to identify hazards, assess risks and opportunities for improvement, implement controls, involve Workers, monitor incidents and near misses, and review and improve the system regularly. This means proactive OH&S is not based on one isolated activity. It depends on a cycle in which the organization looks ahead, plans controls, checks performance, and improves before weaknesses grow into incidents.

In small enterprises, promoting this approach requires visible leadership. ISO's public material identifies leadership and worker participation as key elements of ISO 45001, and its official Q&A states that worker consultation and participation is a requirement of the standard. A proactive workplace therefore depends on leaders who ask about hazards, support reporting, respond to unsafe conditions without delay, and involve Workers in practical safety discussions. When management behaves this way consistently, OH&S becomes part of routine decision-making rather than something addressed only after a problem occurs.

Worker involvement is also essential because proactive OH&S depends on early information. ISO's current guidance specifically recommends identifying hazards and unsafe conditions, involving Workers in safety discussions, monitoring incidents and near misses, and planning corrective and Preventive Emergency arrangements should be communicated clearly to Workers. Workers are often the first to notice weak controls, changing conditions, unsafe shortcuts, equipment problems, or unrealistic procedures. If the organization encourages them to speak up and uses that information constructively, it can act earlier and more effectively. In this way, a proactive approach is closely connected to consultation, participation, communication, and a positive reporting culture.

A proactive approach also means choosing prevention over dependence on last-minute protection. ISO's occupational safety guidance explains that effective OH&S requires smart, tangible preventive measures, and its broader OH&S pages emphasize that even simple changes can prevent serious incidents. In practical small-business terms, this means reviewing tasks before they cause strain or injury, correcting poor housekeeping before it creates a slip or trip hazard, maintaining equipment before failure occurs, planning contractor control before outside work begins, and improving communication before confusion turns into unsafe action. The key idea is that prevention should be built into work, not added only after damage is done.

Another important part of promoting a proactive approach is learning continuously from evidence. ISO links ISO 45001 with monitoring, measurement, performance evaluation, incident investigation, and continual improvement, and its small-organization guidance highlights better incident investigation and stronger opportunities to improve OH&S management as practical benefits of implementation. This means a proactive small enterprise does not simply correct visible problems and move on. It examines reports, incidents, near misses, inspections, audits, and review Observations and recommendations must be documented; implementation is per organization discretion. to understand what they reveal about the strength of the system. Over time, this turns OH&S into a more mature and stable management process.

Promoting a proactive approach ultimately helps create a workplace culture in which health and safety is ingrained in everyday work. ISO's hazards guidance says management system standards such as ISO 45001 can transform how organizations approach risk, helping create a work environment where health and safety is built into everything they do, while ISO's occupational safety guide describes OH&S as creating a culture in which every person can carry out duties without putting life, health, or fitness at risk. For small enterprises, this is the practical goal of proactive OH&S: a workplace where people do not wait for warning signs to become injuries, and where management treats prevention as part of responsible, sustainable business operation.

Verifying the Effectiveness of Corrective Actions

Verifying the effectiveness of corrective actions means checking whether the action taken has actually solved the problem that led to the nonconformity and whether it has reduced the chance of recurrence. In ISO management system language, corrective action is not complete when an issue is recorded or an immediate fix is made. The harmonized ISO management system structure requires the organization to **review the effectiveness of any corrective action taken**, and it also requires documented information as evidence of the results of corrective action.

This is an important distinction because a **correction** and a **corrective action** are not the same. A correction removes the immediate problem that was found, while a corrective action addresses the cause or causes so the issue is less likely to happen again. ISO's harmonized structure guidance explicitly highlights this difference and notes that corrective action should be appropriate to the effects of the nonconformities encountered.

In practical OH&S terms, effectiveness verification asks a simple but important question: **Did the action really work?** If a workplace hazard was reported, an audit raised a finding, or an incident revealed a control failure, management should not assume the issue is closed just because someone signed off an action. The organization needs to confirm that the revised control is in place, understood, used correctly, and producing the intended result in actual work. ISO also links evaluation of system effectiveness to performance evaluation more broadly, which reinforces that follow-up should be based on evidence rather than assumption.

For small enterprises, this step is especially important because resources are limited and repeated failures can quickly damage confidence in the OH&S system. If the same unsafe condition returns, the same audit finding reappears, or the same type of incident happens again, that usually means the earlier corrective action was incomplete, was aimed at the wrong cause, or was not implemented effectively. Verifying effectiveness helps management avoid false closure and directs attention back to the real weakness in the system.

A practical verification process usually checks:

- whether the corrective action was completed as planned
- whether the underlying cause was addressed, not only the visible symptom
- whether the problem has stopped recurring in practice
- whether any related procedures, training, responsibilities, or controls also needed updating
- whether evidence of the result has been retained as documented information

What Effectiveness Verification Looks Like in Practice

In a small enterprise, effectiveness verification should be practical and proportionate. It may involve a follow-up inspection, supervisor observation, review of incident trends, interview with Workers, check of updated instructions, or confirmation that overdue actions and repeated Observations and recommendations must be documented; implementation is per organization discretion. have stopped appearing. The purpose is to confirm that the corrective action has strengthened control, not just changed paperwork. ISO's requirements for monitoring, evaluation, audit, and review all support this evidence-based approach.

For example, if an incident occurred because a machine guard was missing, the immediate correction may be to replace the guard. The corrective action may be broader, such as improving maintenance checks, assigning inspection responsibility more clearly, and retraining supervisors on equipment-release control. Verifying effectiveness would then mean checking not only that the guard is back in place, but also that the new inspection process is working and similar equipment is being controlled properly. That is the difference between fixing one event and improving the OH&S system. This approach is consistent with ISO's requirement to make changes to the management system if necessary after reviewing corrective action effectiveness.

Common Signs That a Corrective Action Was Not Effective

Some warning signs show that a corrective action may not have worked as intended:

- the same nonconformity appears again in a later audit
- a similar incident or near miss happens after the action was closed
- Workers do not understand the new control or do not follow it consistently
- records show completion, but workplace practice has not changed
- the action addressed symptoms, but not the real source of the problem

Table: Verifying the Effectiveness of Corrective Actions

Verification Point	What Should Be Checked	Practical Purpose
Completion of action	Was the corrective action fully implemented as planned?	Confirms the action was not closed early or only partly completed
Cause addressed	Did the action deal with the underlying cause, not only the symptom?	Reduces the chance of recurrence
Workplace result	Has the problem stopped appearing in real operations?	Shows whether control actually improved
Related system changes	Were procedures, responsibilities, training, or documents updated where needed?	Ensures the wider OH&S system supports the fix

Evidence retained	Is there documented information showing the result of the corrective action?	Supports accountability and future review

Why This Matters for OH&S Improvement

Verifying the effectiveness of corrective actions is one of the clearest ways an organization turns nonconformity management into real improvement. It closes the gap between action taken and result achieved. In ISO terms, it supports the suitability, adequacy, and effectiveness of the management system and helps ensure that improvement is based on evidence. For a small enterprise, it also protects time and resources by helping management focus on solutions that actually work.

Strengthening Worker Involvement in Continual Improvement

Strengthening worker involvement in continual improvement means making Workers active contributors to how the OH&S management system is reviewed, corrected, and improved over time. In ISO 45001:2018, this is not treated as an optional extra. ISO identifies **leadership and worker participation** and **continual improvement** as key requirements, and its current guidance says organizations use ISO 45001 by involving Workers in safety processes, monitoring incidents and near misses, and reviewing and improving the system regularly.

Worker involvement matters because continual improvement depends on real information from the workplace. Workers are often the first to notice unsafe conditions, weak controls, unrealistic procedures, equipment problems, recurring shortcuts, or changes in work that increase risk. ISO's practical first steps for ISO 45001 specifically include identifying hazards and unsafe conditions, involving Workers in safety discussions, monitoring incidents and near misses, and planning corrective and Preventive Emergency arrangements should be communicated clearly to Workers. This shows that worker input is part of the improvement process itself, not a separate activity alongside it.

ISO's official Q&A also makes clear that **consultation and participation of Workers is a requirement of ISO 45001**, and that top management must ensure this happens if the organization wants to claim conformity with the standard. That point is especially important in small enterprises, where leaders are usually close to operations and can directly influence whether Workers feel heard, respected, and encouraged to contribute. When Workers are involved properly, improvement becomes more practical because it is based on actual work conditions rather than management assumptions alone.

In practical terms, worker involvement in continual improvement means more than asking for comments occasionally. It means creating regular ways for Workers to contribute to identifying hazards, reviewing incidents and near misses, discussing audit Observations and recommendations must be documented; implementation is per organization discretion., checking whether controls are realistic, suggesting safer work methods, and helping test whether corrective actions have worked. ISO's occupational safety guidance says ISO 45001 helps embed a culture of continuous improvement and empowers Workers to play an active role in creating safer workplaces.

For small enterprises, this involvement should be simple, direct, and consistent. It can happen through short safety discussions, toolbox talks, incident reviews, supervisor conversations, inspection walk-throughs, follow-up after corrective actions, and regular discussion of changes in work methods. The ISO small-organization guide also stresses leadership and worker participation as central to the system and explains the OH&S framework through the PDCA cycle, which means Workers should be involved not only in identifying problems but also in planning, checking, and improving controls.

A strong worker-involvement approach usually includes:

- inviting Workers to raise hazards, concerns, and improvement ideas
- discussing incidents, near misses, and audit Observations and recommendations must be documented; implementation is per organization discretion. openly
- asking whether procedures and controls work in real conditions
- involving Workers in reviewing corrective actions and changes
- giving feedback on what action was taken and what improved

Why It Strengthens Continual Improvement

Continual improvement becomes stronger when worker involvement is real and visible. Without worker input, management may miss early warning signs or keep controls that look good on paper but do not work well in practice. With worker input, the organization is more likely to detect recurring weaknesses, understand the real causes of problems, and choose improvements that are practical to apply. ISO's current guidance describes ISO 45001 as a framework that helps organizations move from reactive safety measures to proactive risk management, and worker participation is one of the mechanisms that makes that shift possible.

Worker involvement also strengthens safety culture. ISO's launch material says ISO 45001 places more focus on leadership and worker participation, with top management expected to take a visible, directing role and ensure integration of the OH&S system into business processes. In a small enterprise, this means Workers are more likely to contribute honestly when leaders respond fairly, act on useful feedback, and show that improvement is part of normal business management.

Area of Worker Involvement	How It Supports Continual Improvement
Hazard and unsafe-condition reporting	Gives early warning of weak controls and new risks
Incident and near-miss discussion	Helps identify causes and prevent recurrence
Feedback on procedures and tasks	Shows whether controls are realistic in daily work
Participation in corrective-action follow-up	Confirms whether changes actually worked
Input during reviews and safety discussions	Keeps improvement aligned with real workplace conditions

In summary, strengthening worker involvement in continual improvement means making Workers part of how the OH&S management system learns, adapts, and gets better. Under ISO 45001, that supports better based on based on risk assessment methods defined by the organization in line with ISO 45001:2018 Clause 6.1.2" to align with risk-based thinking., more realistic controls, stronger corrective action, and a more proactive safety culture over time.

Summary

Verifying the effectiveness of corrective actions means checking whether the action taken has truly solved the problem and reduced the chance of it happening again. It is not enough to close a record or correct the visible issue. The organization must confirm that the real cause was addressed, that the new control is working in practice, and that the same weakness is not returning. This helps small enterprises avoid false closure and ensures that corrective action leads to real improvement.

Strengthening worker involvement in continual improvement means making Workers active participants in identifying hazards, discussing incidents, reviewing controls, and helping improve the OH&S management system over time. Under ISO 45001:2018, worker participation is an essential part of effective OH&S management. When Workers are listened to and involved in improvement, the organization gains better information, more practical solutions, stronger trust, and a more proactive safety culture. Together, these two areas help small enterprises improve OH&S performance in a practical, consistent, and sustainable way.

A Personal Note from the Author

In this final chapter, I wanted to highlight that improvement is where an OH&S management system proves its real value. It is not enough to identify hazards, create procedures, or respond to incidents once they happen. A strong system must learn continuously from what goes wrong, from what nearly goes wrong, and from what Workers and audits reveal over time. My intention was to show that real progress in OH&S comes from honest reporting, careful review, effective corrective action, and a willingness to strengthen controls before problems grow more serious.

I also wanted to emphasize that a proactive safety culture does not develop through paperwork alone. It grows when people are encouraged to speak up, when management responds fairly and consistently, and when improvement becomes part of everyday business practice. For small enterprises, this is especially important because safety culture is often shaped directly by the attitudes and actions of leadership. My hope is that this chapter helps readers see continual improvement not as an extra task, but as an essential part of building a safer, stronger, and more responsible workplace.

Dear Reader,

Thank you for taking the time to read *Practical Strategies for Small Business OH&S Compliance*. This book was written with the clear purpose of helping small enterprises understand that occupational health and safety is not only a formal requirement, but a daily responsibility that protects people, strengthens operations, and supports long-term business stability.

Throughout these chapters, the goal has been to present ISO 45001:2018 in a practical and accessible way for small business environments. Small enterprises often work under real pressures, limited resources, and constantly changing conditions, yet they still have the ability to build safer, healthier, and more responsible workplaces when OH&S is approached with clarity, commitment, and consistency.

I hope this book has helped you see that effective OH&S compliance is not achieved through paperwork alone. It is built through leadership, worker involvement, practical controls, honest reporting, regular review, and continual improvement. Even small steps, when taken seriously and applied consistently, can make a meaningful difference in reducing risk and improving workplace safety performance.

As you move forward, I encourage you to treat occupational health and safety not as a one-time effort, but as an ongoing part of responsible management and professional practice. Safer workplaces are created when commitment is translated into daily action.

Kind regards,
Dr. Mohamed Ali Ibrahim
Vienna, Austria

PART 4

ISO 45001 Documentation Kit for Small Enterprises

Practical Templates and Tools for Implementation, Compliance, and Audit Readiness

CHAPTER 16

OH&S Policy and System Foundation Templates

OH&S Policy Template

Instructions for Use

Use this template to create an OH&S Policy that reflects the nature of your organization, the risks arising from your activities, and your commitment to maintaining a safe and healthy workplace. The policy should be approved by top management, communicated within the organization, kept as documented information, and reviewed when necessary to ensure it remains relevant to the business.

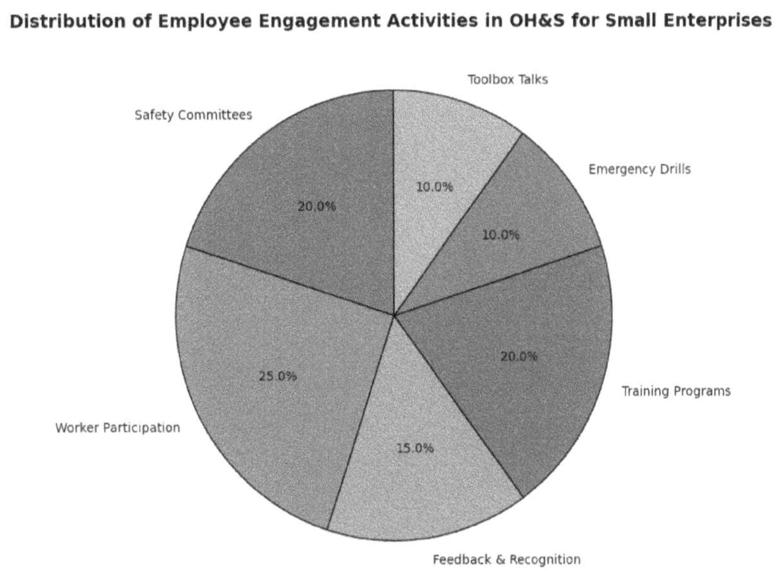

Occupational Health & Safety (OH&S) Policy Template for Small Enterprises

Company Name: ____________________________
Address: ___________________________________
Contact Email: _____________________________
Phone Number: _____________________________

---

1. Policy Statement

[Your Company Name] is committed to ensuring the health, safety, and welfare of all employees and any other persons who may be affected by our operations. Our goal is to provide a safe and healthy work environment and to prevent accidents, injuries, and work-related illnesses.

---

2. Responsibilities

2.1 Management Responsibilities

- Ensure that all OH&S policies are implemented, maintained, and communicated to all employees.
- Provide necessary resources to support and enforce these policies.
- Regularly review and update OH&S policies and procedures.

2.2 Employee Responsibilities

- Adhere to all health and safety procedures and policies.
- Report any unsafe conditions or practices to management.
- Participate in safety training, drills, and consultations.

---

3. Risk Assessments

- Regular risk assessments will be conducted to identify potential hazards.
- Mitigation strategies will be developed and implemented to control identified risks.

---

4. Training and Awareness

- All employees will receive OH&S training relevant to their roles.
- Ongoing training will be provided to ensure employees remain up-to-date with current health and safety standards.

---

5. Accident Reporting and Investigation

- All accidents and near-misses must be reported immediately to management.
- Investigations will be conducted to determine the cause and prevent future occurrences.

---

6. Emergency Procedures

- Emergency procedures, including evacuation plans, will be established and communicated to all employees.
- Regular drills will be conducted to ensure readiness in case of emergencies.

---

7. Health and Safety Monitoring

- Regular inspections and audits will be conducted to ensure compliance with health and safety policies.
- Non-compliance issues will be addressed promptly.

---

8. Review and Continuous Improvement

- This policy will be reviewed annually and updated as necessary to reflect changes in legislation, industry standards, and company operations.

---

9. Declaration

By adhering to this policy, [Your Company Name] demonstrates its commitment to maintaining a safe and healthy work environment for all.

Signature (Top Management): ______________________
Date: ______________________

OH&S Management System Scope Statement Template

Company Name: ______________________________
Address: _____________________________________
Contact Email: ______________________________
Phone Number: ______________________________

---

1. Scope Statement Overview

This Scope Statement defines the boundaries and applicability of the Occupational Health and Safety (OH&S) Management System for [Your Company Name]. It establishes which activities, locations, and personnel are included within the OH&S Management System and ensures alignment with ISO 45001:2018 requirements.

---

2. Scope Description

Scope Title: ___________________________________

Scope Statement:

Locations Covered:

- ______________________________
- ______________________________
- ______________________________

Departments / Activities Covered:

- ______________________________
- ______________________________
- ______________________________

Personnel Covered:

- ______________________________
- ______________________________
- ______________________________

---

3. Exclusions

List any areas, activities, or personnel excluded from the OH&S Management System scope and provide justification:

- ______________________________
- ______________________________
- ______________________________

---

4. Boundaries

Define the operational and organizational boundaries for the OH&S system:

- ______________________________
- ______________________________
- ______________________________

---

5. Applicable Legal and Other Requirements

Identify key OH&S regulations, client requirements, or industry standards applicable within the defined scope:

- ______________________________
- ______________________________
- ______________________________

6. Review and Approval

This Scope Statement will be reviewed annually or when changes to the organization, operations, or applicable legal requirements occur.

Approved by (Top Management): ______________________
Position: ______________________
Date: ______________________

OH&S Roles and Responsibilities Matrix Template – Detailed Version

Company Name: ______________________

Document: OH&S Roles & Responsibilities Matrix

ISO 45001:2018

---

1. Purpose

This matrix defines all key OH&S roles within [Your Company Name], clarifying responsibilities, accountability, and reporting lines. It ensures compliance with ISO 45001:2018 and supports effective implementation of the OH&S Management System.

---

2. Roles and Responsibilities Matrix

Role / Position	Key Responsibilities	Authority / Decision-Making	Reporting / Accountable To	Notes / Additional Information
Top Management / Owner	Approve OH&S policy, allocate resources, review performance, lead safety culture	Full authority to enforce OH&S system	Board / Self	Overall accountability for OH&S system
OH&S Coordinator / Safety Officer	Maintain OH&S documentation, conduct inspections, track incidents,	Can advise and implement operational controls	Top Management	Primary operational OH&S role

	coordinate training, ensure compliance			
Supervisor / Team Lead	Implement procedures, monitor safe work, conduct toolbox talks, report hazards	Authority to enforce safety procedures in their team	OH&S Coordinator / Top Management	Day-to-day enforcement of OH&S requirements
Employees / Workers	Follow procedures, use PPE, report hazards, participate in training	Compliance-focused	Supervisor / OH&S Coordinator	Direct engagement in operational safety practices
Contractors / Temporary Staff	Follow site-specific OH&S rules, report hazards, attend induction	Compliance-focused	Supervisor / OH&S Coordinator	Must be briefed on site-specific hazards
Emergency Response Team / First Aid Officer	Respond to incidents, maintain emergency equipment, support evacuation	Authority during emergencies	Supervisor / Top Management	Critical during emergencies; requires specialized training
Maintenance Personnel	Maintain equipment, ensure safety before operation, report unsafe equipment	Authority to stop unsafe equipment use	Supervisor / OH&S Coordinator	Works closely with operations for hazard control
Training Coordinator / HR	Organize training, maintain records, evaluate competency	Can schedule and allocate resources	OH&S Coordinator / Top Management	Ensures workforce competence
Internal Audit Lead	Conduct internal audits, assess compliance, recommend improvements	Can recommend corrective actions	OH&S Coordinator / Top Management	Ensures OH&S system effectiveness

3. Guidelines for Using the Matrix

- Ensure each responsibility is clearly assigned.
- Update the matrix whenever there are organizational, operational, or regulatory changes.

- Share the matrix with all staff during induction and make it accessible for reference.
- Maintain documentation linked to each role for auditing purposes.
- Integrate the matrix into internal audits, management reviews, and ISO 45001 certification processes.

Page: ___ of ___ | **Confidential – Internal Use Only** | **Version:** ___

OH&S Objectives Template – Small Enterprises

Company Name: ____________________________

Document: OH&S Objectives

ISO 45001:2018

Company Information

- **Address:** __________________________________
- **Contact Email:** ____________________________
- **Phone Number:** ____________________________

---

1. Purpose

This template defines the **Occupational Health and Safety objectives** for [Your Company Name]. OH&S objectives provide measurable targets to reduce risk, improve safety performance, and ensure compliance with ISO 45001:2018. Objectives should be specific, measurable, achievable, relevant, and time-bound (SMART).

---

2. OH&S Objectives Table

Objective	Description / Scope	Target / Metric	Responsible Person	Resources Required	Target Completion Date	Status / Notes
Example: Reduce workplace incidents	Decrease slips, trips, and falls in the warehouse	<2 incidents per quarter	Warehouse Supervisor	Training, signage, housekeeping	31/12/2026	Ongoing

3. Guidelines for Setting OH&S Objectives

- **SMART Objectives:** Ensure each objective is Specific, Measurable, Achievable, Relevant, and Time-bound.
- **Link to Risks:** Align objectives with identified hazards and operational risks.
- **Assign Responsibility:** Clearly identify the person accountable for achieving each objective.
- **Allocate Resources:** Ensure necessary resources, including training, equipment, or budget, are available.
- **Monitor and Review:** Track progress regularly and update objectives as needed to reflect changes in operations or regulations.

CHAPTER 17

Based on based on risk assessment methods defined by the organization in line with ISO 45001:2018 Clause 6.1.2" to align with risk-based thinking. and Risk Control Templates

In any Occupational Health and Safety (OH&S) Management System, the ability to systematically **identify hazards and control associated risks** is fundamental to protecting employees and ensuring compliance with ISO 45001:2018. For small enterprises, where operations are often fast-moving and employees may perform multiple roles, having **structured templates** makes based on based on risk assessment methods defined by the organization in line with ISO 45001:2018 Clause 6.1.2" to align with risk-based thinking. and risk control more practical, consistent, and auditable.

This chapter provides **ready-to-use templates** that allow small businesses to record hazards, assess risks, document control measures, assign responsibilities, and track the effectiveness of mitigation strategies. By using these templates, organizations can standardize their approach, maintain clear records, and demonstrate due diligence to auditors, regulatory bodies, and stakeholders.

The templates are designed to be **fillable, editable, and aligned with ISO 45001:2018** requirements, ensuring that small enterprises can implement them efficiently without unnecessary complexity. Each template includes professional **headers and footers**, safe margins, and structured fields for clarity.

These templates serve as the backbone for proactive risk management, helping small enterprises:

- Identify workplace hazards systematically
- Assess the likelihood and severity of risks
- Define appropriate control measures
- Assign responsibilities and accountability
- Monitor and review risk controls continuously

The following sections will provide **practical, ready-to-use templates** for based on based on risk assessment methods defined by the organization in line with ISO 45001:2018 Clause 6.1.2" to align with risk-based thinking., risk assessment, and risk control suitable for small enterprises across all business sectors.

Checklist Template – Small Enterprises

Company Name: ____________________________ | **Document:** Based on based on risk assessment methods defined by the organization in line with ISO 45001:2018 Clause 6.1.2" to align with risk-based thinking. Checklist | **ISO 45001:2018**

Company Information

- **Address:** __________________________________
- **Contact Email:** _____________________________
- **Phone Number:** _____________________________

---

1. Purpose

This checklist is intended to systematically identify hazards in the workplace, evaluate risks, and document the necessary control measures. Using this checklist supports small enterprises in maintaining a safe work environment, ensuring ISO 45001:2018 compliance, and preparing for internal or external audits.

---

2. Based on based on risk assessment methods defined by the organization in line with ISO 45001:2018 Clause 6.1.2" to align with risk-based thinking. Checklist Table

Area / Task	Potential Hazard	Risk Level (High/Medium/Low)	Existing Controls	Additional Controls Required	Responsible Person	Date Completed	Notes
Example: Warehouse – Stock handling	Manual handling injuries	High	Lifting training, PPE	Introduce lifting aids	Warehouse Supervisor	01/03/2026	Reviewed quarterly
_______ _______	_______ _______	_______ _______	_______ _______	_______ _______	_______ _______	_______ _______	_______ _______

________ ______	________ ______	________ ______	________ ______	________ ______	________ ___	________ ___	________ ______
________ ________ ________ ______	________ ________ ________ ______	________ ________ ________ ______	________ ________ ________ ______	________ ________ ________ ______	________ ________ ________ ___	________ ________ ________ ___	________ ________ ________ ______
________ ________ ________ ______	________ ________ ________ ______	________ ________ ________ ______	________ ________ ________ ______	________ ________ ________ ______	________ ________ ________ ___	________ ________ ________ ___	________ ________ ________ ______
________ ________ ________ ______	________ ________ ________ ______	________ ________ ________ ______	________ ________ ________ ______	________ ________ ________ ______	________ ________ ________ ___	________ ________ ________ ___	________ ________ ________ ______

3. Instructions for Use

- Inspect all areas, tasks, and processes systematically.
- Identify potential hazards related to physical, chemical, biological, ergonomic, and psychosocial risks.
- Assess risk level based on likelihood and severity.
- Record existing controls and identify additional control measures if required.
- Assign responsibility for implementing controls.
- Complete the checklist and review regularly to maintain a proactive OH&S system.

Page: ___ of ___ | **Confidential – Internal Use Only** | **Version:** ___ | **Year:** 20__

Risk Assessment Matrix Template – Small Enterprises

Company Name: ____________________________ | **Document:** Risk Assessment Matrix | **ISO 45001:2018**

Company Information

- **Address:** ___________________________________
- **Contact Email:** ______________________________
- **Phone Number:** ______________________________

---

1. Purpose

The Risk Assessment Matrix Template is used to **systematically evaluate workplace hazards**, determine the likelihood and severity of each risk, assign a risk rating, and identify suitable control measures. It ensures small enterprises manage risks proactively and maintain compliance with ISO 45001:2018.

---

2. Risk Assessment Matrix Table

Area / Task	Hazard Identified	Likelihood (1–5)	Severity (1–5)	Risk Rating (L x S)	Existing Controls	Additional Controls Required	Responsible Person	Target Completion Date	Notes
____________________	____________________	____________________	____________________	____________________	____________________	____________________	____________________	____________________	______

3. Instructions for Using the Risk Assessment Matrix

- **Step 1 – Identify Area/Task:** Define the workplace location, task, or activity being assessed.
- **Step 2 – Identify Hazard:** Record all potential hazards (physical, chemical, biological, ergonomic, psychosocial, or operational).

- **Step 3 – Likelihood and Severity:** Rate the likelihood of occurrence (1 = Rare, 5 = Almost Certain) and severity of consequences (1 = Minor, 5 = Major/Fatal).
- **Step 4 – Calculate Risk Rating:** Multiply Likelihood x Severity to assign the risk level. Example: 4 x 4 = 16 (High).
- **Step 5 – Existing Controls:** Document controls already in place to reduce the risk.
- **Step 6 – Additional Controls:** Identify actions required to further mitigate risk.
- **Step 7 – Responsible Person:** Assign accountability for implementing additional controls.
- **Step 8 – Target Completion Date:** Set a realistic date for implementing controls.
- **Step 9 – Notes:** Include observations, review dates, or comments on risk changes.

---

4. Risk Level Guidelines

Risk Rating (L x S)	Risk Level	Recommended Action
1–5	Low	Monitor; maintain existing controls
6–10	Medium	Apply additional controls where feasible
11–15	High	Implement additional controls immediately; review frequently
16–25	Critical	Stop work if necessary; implement controls urgently; escalate to management

Footer (appears on every page):
Page: ___ of ___ | **Confidential – Internal Use Only** | **Version:** ___ | **Year:** 20__

Risk Control / Action Plan Template – Small Enterprises

Company Name: ____________________________ | **Document:** Risk Control / Action Plan | **ISO 45001:2018**

Company Information

- **Address:** ______________________________
- **Contact Email:** _____________________________
- **Phone Number:** _____________________________

---

1. Purpose

This Risk Control / Action Plan Template is designed to track **actions taken to control identified hazards and mitigate risks**. It ensures responsibilities, timelines, and control measures are clearly assigned and documented for audit readiness and ISO 45001:2018 compliance.

---

2. Risk Control / Action Plan Table

Hazard / Risk	Control Measure	Responsible Person	Target Completion Date	Status
Example: Manual handling injury	Provide lifting training & PPE	Warehouse Supervisor	30/06/2026	Ongoing
Slips / Trips	Improve floor signage & housekeeping	Supervisor	15/05/2026	Not Started
Machine entanglement	Install safety guards	Maintenance Lead	10/06/2026	In Progress
Chemical exposure	Provide gloves & safety data	Safety Officer	20/05/2026	Completed
Fire risk	Conduct fire drill & check extinguishers	HR / Safety Officer	05/06/2026	Scheduled

Electrical hazard	Inspect wiring & lockout/tagout	Maintenance Lead	12/06/2026	In Progress
Noise exposure	Provide ear protection & training	Supervisor	25/05/2026	Ongoing

3. Guidelines for Using the Template

- List hazards or risks identified in the OH&S system.
- Specify clear **control measures** for each risk.
- Assign **responsible person(s)** to implement the control.
- Set **target completion dates** for implementation.
- Track **status** (e.g., Not Started, In Progress, Completed, Ongoing).
- Update regularly and review during audits or management reviews.

Footer (appears on every page):
Page: ___ of ___ | **Confidential – Internal Use Only** | **Version:** ___ | **Year:** 20__

CHAPTER 18

Competence, Training, and Awareness Templates

This chapter provides **ready-to-use templates** to help small enterprises document, track, and manage **employee competence, training, and awareness** in accordance with ISO 45001:2018. These templates are designed to simplify planning, record-keeping, and verification, ensuring that all staff have the necessary skills and knowledge to perform their roles safely.

By using these templates, small businesses can:

- Identify training needs and skill gaps efficiently
- Record completed training and certifications
- Track employee awareness of hazards and safe practices
- Ensure ongoing competency for operational safety
- Demonstrate compliance during audits and internal reviews

These templates are **fillable, editable, and structured with headers and footers**, making them practical for everyday use while maintaining ISO 45001 compliance.

Competence and Training Matrix Template – Small Enterprises

Company Name: ___________________________ | **Document:** Yes/No indicates verified demonstration of competency; additional notes may be required for clarification | **ISO 45001:2018**

Company Information

- **Address:** _________________________________
- **Contact Email:** ____________________________
- **Phone Number:** ____________________________

---

1. Purpose

This Competence and Training Matrix Template helps small enterprises systematically document **employee skills, qualifications, and OH&S training**, ensuring that staff are competent for their roles and aware of workplace hazards. It supports ISO 45001:2018 compliance, continuous improvement, and audit readiness.

---

2. Yes/No indicates verified demonstration of competency; additional notes may be required for clarification Table

Employee Name	Job Role	Required Competence / Training	Training Completed (Yes/No)	Date Completed	Next Refresher Date	Responsible Person	Notes / Comments
John Smith	Warehouse Operator	Manual handling, Forklift operation, PPE use	Yes	15/03/2026	15/03/2027	Supervisor	Competent in all tasks

Jane Doe	Maintenance Technician	Machine safety, Lockout/Tagout, Emergency response	No	-	01/06/2026	Safety Officer	Scheduled for training
Mark Johnson	Office Staff	Ergonomics, Emergency evacuation, Fire safety	Yes	01/02/2026	01/02/2027	HR Manager	Refresher due next year
Contractor A	External Electrician	Electrical safety, PPE, Hazard awareness	Yes	10/03/2026	10/03/2027	Supervisor	Works under supervision
________	________	________	________	________	________	________	________
________	________	________	________	________	________	________	________

---

3. Instructions for Use

- List all employees, contractors, and temporary staff.
- Identify **required competencies and training** for each role.
- Record completion of training and certification dates.
- Schedule next refresher dates as required.
- Assign responsibility for ensuring training completion.
- Include additional notes on skill gaps or observations.
- Update regularly to maintain accuracy and support audits.

Footer (appears on every page):
Page: ___ of ___ | **Confidential – Internal Use Only** | **Version:** ___ | **Year:** 20__

Employee Training Log Template – Small Enterprises

Company Name: ______________________ | **Document:** Employee Training Log | **ISO 45001:2018**

Company Information

- **Address:** ____________________________
- **Contact Email:** ______________________
- **Phone Number:** _______________________

---

1. Purpose

This template records **all employee OH&S training activities** in a concise, structured format. Single-word table entries make it easier to fill, print, and track completion while supporting ISO 45001 audits.

---

2. Employee Training Log Table

Employee	Role	Topic	Type	Trainer	Date	Next	Status	Notes
John	Warehouse	Lifting	Refresher	Supervisor	15/03/26	15/03/27	Yes	Good
Jane	Maintenance	Machine	Task	Safety	10/03/26	10/03/27	Yes	Review
Mark	Office	Fire	Induction	HR	01/02/26	01/02/27	Yes	Due
Contractor	Electrician	Electrical	Induction	Supervisor	10/03/26	10/03/27	Yes	Supervised
________	________	________	________	________	________	________	________	________

3. Instructions for Use

- Fill in **Employee, Role, Topic, Type, Trainer, Date, Next, Status, Notes**.
- Use **single-word entries** to ensure clarity and compact layout.
- Update regularly to reflect completed training and schedule future sessions.
- Keep as a record for ISO 45001 compliance and audits.

Footer (appears on every page):
Page: ___ of ___ | **Confidential** | **Version:** ___ | **Year:** 20__

Competency Assessment Template – Small Enterprises

Company Name: ___________________________ | **Document:** Competency Assessment | **ISO 45001:2018**

Company Information

- **Address:** ______________________________
- **Contact Email:** __________________________
- **Phone Number:** __________________________

---

1. Purpose

This Competency Assessment Template allows small enterprises to **evaluate employee knowledge, skills, and abilities** for specific OH&S tasks. It supports ongoing training, performance monitoring, and ISO 45001:2018 compliance.

---

2. Competency Assessment Table

Employee	Role	Task	Knowledge	Skill	Performance	Assessment Date	Evaluator	Status
John	Warehouse	Lifting	Yes	Yes	Pass	15/03/26	Supervisor	Competent
Jane	Maintenance	Machine	Yes	Yes	Pass	10/03/26	Safety	Competent
Mark	Office	Fire	Yes	No	Fail	01/02/26	HR	NeedsReview
Contractor	Electrician	Electrical	Yes	Yes	Pass	10/03/26	Supervisor	Competent
______	______	______	______	______	______	______	______	______

3. Instructions for Use

- Evaluate each employee for **Knowledge**, **Skill**, and **Performance** on specific tasks.
- Use single-word entries such as **Yes/No, Pass/Fail, Competent/NeedsReview**.
- Record the **assessment date** and **evaluator** for audit traceability.
- Update assessments regularly to maintain competency records and plan further training.
- Include this document in ISO 45001:2018 compliance records.

Footer (appears on every page):
Page: ___ of ___ | **Confidential – Internal Use Only** | **Version:** ___ | **Year:** 20__

Training Evaluation Template – Small Enterprises

Company Name: ____________________________ | **Document:** Training Evaluation Form | **ISO 45001:2018**

Company Information

- **Address:** ___________________________________
- **Contact Email:** _____________________________
- **Phone Number:** _____________________________

---

1. Purpose

This Training Evaluation Template is used to **assess the effectiveness of OH&S training** within small enterprises. It allows management to measure whether employees have acquired the required knowledge and skills, identify gaps, and plan follow-up actions. The template supports ISO 45001:2018 compliance, audit readiness, and continuous improvement of training programs.

2. Training Evaluation Table

Employee	Role	Training Topic	Date	Trainer	Knowledge Gained (Yes/No)	Confidence Level (High/Medium/Low)		Comments
______	___ ___	_____ _	___ ___	____ _	_____	_____		______

---

3. Instructions for Using the Training Evaluation Template

- Fill in **Employee, Role, Training Topic, Date, and Trainer**.
- Evaluate **Knowledge Gained**: whether the employee understood the training content.
- Evaluate **Skill Applied**: whether the employee can perform the task safely and correctly.
- Record **Confidence Level**: High, Medium, or Low.
- Note if **Follow-up Training** is required.
- Use the **Comments** section for observations, improvement recommendations, or corrective action.
- Maintain completed evaluations as part of ISO 45001:2018 documentation for audits and continuous improvement.

Footer (appears on every page):
Page: ___ of ___ | **Confidential – Internal Use Only** | **Version:** ___ | **Year:** 20__

CHAPTER 19

Operational Control and Improvement Templates

Evaluating training effectiveness is a critical part of any **Occupational Health and Safety (OH&S) Management System**, particularly for small enterprises where staff often perform multiple roles and work under dynamic conditions. Simply delivering training is not enough; organizations must assess whether employees have **acquired the required knowledge, developed the necessary skills, and can apply them safely in the workplace**.

This chapter provides **fillable, structured templates** to document and evaluate OH&S training outcomes. These templates allow small enterprises to:

- Track **knowledge and skill acquisition** for each employee
- Measure **confidence and competency** in performing tasks safely
- Identify **training gaps** and schedule follow-up sessions
- Maintain records for **ISO 45001:2018 compliance** and audits
- Support **continuous improvement** of the OH&S training program

The templates are designed to be **editable, professional, and easy to use**, with **headers, footers, safe margins, and compact tables**, ensuring that all relevant information is captured in a clear and organized format.

By using these templates, small enterprises can systematically evaluate the effectiveness of their training programs, enhance workforce competency, and strengthen their overall safety culture.

Nonconformity Report Template – Small Enterprises

Company: ___________________ | **Document:** Nonconformity Report | **ISO 45001:2018**

Company Info:

- Address: ____________________________
- Email: ______________________________
- Phone: _____________________________

---

1. Purpose

This template records **all nonconformities**, tracks corrective actions, and monitors verification. Designed for quick, concise entries suitable for small enterprises.

---

2. Nonconformity Details

ID	Date	Dept	Hazard	Severity	Action	Reported By
NCR-001	15/03	Warehouse	Lifting	Med	PPE	John
NCR-002	10/03	Maint	Guard	High	Stop	Jane
NCR-003	01/02	Office	Exit	Med	Clear	Mark
______	______	______	______	______	______	______

---

3. Investigation & Corrective Action

ID	Cause	Corrective	Resp	Due	Status
NCR-001	Forgot	Training	Sup	20/03	Ong
NCR-002	Maint	Repair	Maint	18/03	Done
NCR-003	House	Inspect	HR	15/03	Ong
______	______	______	______	______	______

---

4. Verification

ID	Verified By	Date	Effect	Notes
NCR-001	Safety	21/03	Yes	Good
NCR-002	Sup	19/03	Yes	Ok
NCR-003	HR	16/03	Yes	Ok
______	______	______	______	______

---

5. Instructions

- Use short words in each table cell.
- Assign **unique IDs** for all nonconformities.
- Track **department, hazard, severity, corrective action, and responsible person**.
- Monitor **status** and verify corrective actions.
- Maintain for **ISO 45001 audits** and internal reviews.

Footer (every page):
Page: ___ of ___ | **Confidential** | **Version:** ___ | **Year:** 20__

Corrective Action Request (CAR) Template – Small Enterprises

Company: ____________________ | **Document:** Corrective Action Request | **ISO 45001:2018**

Company Info:

- Address: ____________________________
- Email: ______________________________
- Phone: _____________________________

---

1. Purpose

This template documents **corrective actions** taken to address nonconformities, incidents, or hazards. It helps small enterprises track **responsibility, deadlines, and effectiveness**, supporting ISO 45001:2018 compliance and audit readiness.

---

2. Corrective Action Request Table

CAR ID	Date	Dept	Issue	Cause	Action	Resp	Due	Status
CAR-001	15/03	Warehouse	PPE	Forgot	Train	Sup	20/03	Ong
CAR-002	10/03	Maint	Guard	Maint	Repair	Maint	18/03	Done
CAR-003	01/02	Office	Exit	House	Inspect	HR	15/03	Ong
______	______	______	______	______	______	______	______	______

3. Verification Table

CAR ID	Verified By	Date	Effect	Notes
CAR-001	Safety	21/03	Yes	Good
CAR-002	Sup	19/03	Yes	Ok
CAR-003	HR	16/03	Yes	Ok

_____	_____	_____	_____	_____

4. Instructions for Use

- Assign **unique CAR IDs** for each corrective action.
- Record **department, issue, cause, action, responsible person, due date, and status**.
- Verify completion and **effectiveness** of each corrective action.
- Maintain records for **audit purposes** and ISO 45001 compliance.
- Use **single-word or short entries** to fit compact table cells.

Page: ___ of ___ | **Confidential** | **Version:** ___ | **Year:** 20__

Root Cause Analysis (RCA) Template – Small Enterprises

Company: ___________________ | **Document:** Root Cause Analysis | **ISO 45001:2018**

Company Info:

- Address: ____________________________
- Email: ______________________________
- Phone: _____________________________

---

1. Purpose

This template provides a structured format to **investigate nonconformities, incidents, or hazards**, determine their root causes, and track corrective actions. It ensures small enterprises can address the source of issues and prevent recurrence while remaining ISO 45001:2018 compliant.

---

2. Incident / Nonconformity Details Table

RCA ID	Date	Dept	Issue	Severity	Reported By	Immediate Action
RCA-001	15/03	Warehouse	PPE Missing	Med	John	Issued PPE
RCA-002	10/03	Maint	Guard Missing	High	Jane	Machine Stop
RCA-003	01/02	Office	Exit Blocked	Med	Mark	Cleared Exit
______	______	______	______	______	______	______

3. Root Cause Analysis Table

RCA ID	Root Cause	Corrective Action	Responsible	Due Date	Status
RCA-001	Forgot	Training	Sup	20/03	Ong
RCA-002	Maint	Repair	Maint	18/03	Done

RCA-003	Housekeeping	Weekly Check	HR	15/03	Ong
______	______	______	______	______	______

4. Verification Table

RCA ID	Verified By	Verification Date	Effectiveness (Yes/No)	Notes
RCA-001	Safety	21/03	Yes	Good
RCA-002	Sup	19/03	Yes	Ok
RCA-003	HR	16/03	Yes	Ok
______	______	______	______	______

---

5. Instructions for Use

- Assign **unique RCA IDs** for each incident or nonconformity.
- Record **department, issue, severity, reporter, and immediate action**.
- Identify **root cause**, assign **corrective action**, **responsible person**, and **due date**.
- Verify completion and **effectiveness** using the verification table.
- Maintain records for ISO 45001 compliance, internal audits, and continual improvement.
- Use **short or single-word entries** to fit into table cells.

Page: ___ of ___ | **Confidential – Internal Use Only** | **Version:** ___ | **Year:** 20__

Incident / Near-Miss Report Template – Small Enterprises

Company: ___________________ | **Document:** Incident / Near-Miss Report | **ISO 45001:2018**

Company Information

- Address: ____________________________
- Email: ______________________________
- Phone: _____________________________

---

1. Purpose

This template records **incidents and near-misses** in the workplace. It ensures small enterprises can track events, investigate causes, implement corrective actions, and comply with ISO 45001:2018.

---

2. Incident / Near-Miss Details Table

Report ID	Date	Dept	Type (Incident/NM)	Severity	Reported By	Immediate Action
IN-001	15/03	Warehouse	Incident	Med	John	PPE Issued
IN-002	10/03	Maint	Near-Miss	High	Jane	Machine Stop
IN-003	01/02	Office	Incident	Low	Mark	Exit Cleared
______	______	______	______	______	______	______

---

3. Investigation Table

Report ID	Root Cause	Corrective Action	Responsible	Due Date	Status
IN-001	Forgot	Train	Sup	20/03	Ong
IN-002	Maint	Repair	Maint	18/03	Done

IN-003	Housekeeping	Weekly Check	HR	15/03	Ong
______	______	______	______	______	______

4. Verification Table

Report ID	Verified By	Date	Effectiveness (Yes/No)	Comments
IN-001	Safety	21/03	Yes	Good
IN-002	Sup	19/03	Yes	Ok
IN-003	HR	16/03	Yes	Ok
______	______	______	______	______

5. Instructions for Use

- Assign **unique Report IDs** to each incident or near-miss.
- Record **department, type, severity, reporter, and immediate action**.
- Investigate root causes and define **corrective actions, responsibility, and due dates**.
- Verify corrective actions and record **effectiveness**.
- Keep as a record for ISO 45001 audits and internal reviews.
- Use **short or single-word entries** for table cells.

Footer (appears on every page):
Page: ___ of ___ | **Confidential – Internal Use Only** | **Version:** ___ | **Year:** 20__

CHAPTER 20

Audit and Compliance Readiness Templates

Audits and compliance checks are a critical component of any **Occupational Health and Safety (OH&S) Management System**, particularly under ISO 45001:2018. For small enterprises, preparing for audits ensures that procedures are followed, hazards are controlled, and records are complete and accurate. Proper preparation not only demonstrates compliance but also strengthens the safety culture and operational efficiency of the organization.

This chapter provides **ready-to-use, fillable templates** designed to help small enterprises **document, monitor, and verify** their OH&S compliance in a structured manner. These templates enable organizations to:

- Track internal audits and inspections
- Verify compliance with OH&S procedures, legal requirements, and ISO 45001 standards
- Assign responsibilities for audit tasks and corrective actions
- Record Observations and recommendations must be documented; implementation is per organization discretion., observations, and improvement opportunities
- Prepare for external certification audits and regulatory inspections

The templates are **fillable, professional, and include headers and footers**, ensuring consistent documentation and audit readiness. By using these templates, small enterprises can approach audits with confidence, maintain ISO 45001 compliance, and continuously improve their OH&S management system.

Pre-Audit Checklist Template – Small Enterprises

Company: ________________ | **Document:** Pre-Audit Checklist | **ISO 45001:2018**

Company Information

- Address: ________________________
- Email: __________________________
- Phone: _________________________

---

1. Purpose

This Pre-Audit Checklist Template allows small enterprises to **systematically verify preparedness** before internal or external audits. It ensures documentation, processes, and compliance measures are reviewed and ready for inspection.

---

2. Pre-Audit Checklist Table

Item	Area / Process	Checked (Yes/No)	Responsible Person	Comments
1	OH&S Policy	Yes	Manager	Updated
2	Risk Assessment	Yes	Safety Officer	Reviewed
3	Training Records	No	HR	Schedule
4	Incident Reports	Yes	Supervisor	Completed
5	Corrective Actions	No	Supervisor	Pending
6	Legal Requirements	Yes	Safety Officer	Verified
7	Emergency Procedures	Yes	HR	Drills done
8	Internal Audits	No	Auditor	Planned
9	Equipment Maintenance	Yes	Maintenance Lead	Checked
10	PPE Availability	Yes	Supervisor	Available
______	______	______	______	______

3. Instructions for Use

- List all key areas, processes, and documents to be reviewed before the audit.
- Mark **Checked** as Yes/No to indicate readiness.
- Assign **Responsible Person** for verification of each item.
- Use **Comments** for notes, pending actions, or observations.
- Update the checklist regularly to maintain continuous audit readiness.
- Keep completed checklist as **evidence for internal and external audits**.

Footer (every page):
Page: ___ of ___ | **Confidential – Internal Use Only** | **Version:** ___ | **Year:** 20__

Internal Audit Plan and Report Template – Small Enterprises

Company: ________________ | **Document:** Internal Audit Plan & Report | **ISO 45001:2018**

Company Information

- Address: ________________________
- Email: __________________________
- Phone: _________________________

---

1. Purpose

This Internal Audit Plan and Report Template helps small enterprises **plan, conduct, and document internal OH&S audits**. It ensures compliance with ISO 45001:2018, identifies gaps, and tracks corrective actions for continuous improvement.

---

2. Audit Plan Table

Audit ID	Area / Process	Auditor	Scheduled Date	Scope	Status (Planned/Completed)
AUD-001	Warehouse	John	20/04/26	Risk Assess	Planned
AUD-002	Maintenance	Jane	22/04/26	Equipment	Planned
AUD-003	Office	Mark	25/04/26	Training	Planned
______	______	______	______	______	______

---

3. Audit Observations and recommendations must be documented; implementation is per organization discretion. Table

Audit ID	Observation / Finding	Severity (Low/Med/High)	Nonconformity (Yes/No)	Corrective Action	Responsible	Due Date	Status
AUD-001	PPE not used	Med	Yes	Training	Supervisor	30/04/26	Ong
AUD-002	Guard missing	High	Yes	Repair	Maint	28/04/26	Done
AUD-003	Fire drill overdue	Low	No	Schedule	HR	29/04/26	Ong
______	______	______	______	______	______	______	______

---

4. Instructions for Use

1. Assign **unique Audit IDs** for each planned internal audit.
2. Schedule audits with **auditor, area/process, and scope**.
3. Record **Observations and recommendations must be documented; implementation is per organization discretion., severity, nonconformities, corrective actions, responsible person, due date, and status**.
4. Use **short or single-word entries** for compact table cells.
5. Update the template after audits to maintain accurate records.
6. Keep completed templates as **evidence for ISO 45001 internal and external audits**.

Footer (appears on every page):
Page: ___ of ___ | **Confidential – Internal Use Only** | **Version:** ___ | **Year:** 20__

Legal and Other Requirements Register / Compliance Matrix Template – Small Enterprises

Company: __________________ | **Document:** Legal & Compliance Register | **ISO 45001:2018**

Company Information

- Address: __________________________
- Email: ____________________________
- Phone: ___________________________

---

1. Purpose

This template allows small enterprises to **identify, record, and track applicable legal and other OH&S requirements**, ensuring that responsibilities, compliance status, and review dates are clearly documented in line with ISO 45001:2018.

---

2. Legal & Compliance Register Table

Requirement ID	Source / Law	Area / Process	Applicable (Yes/No)	Responsible	Control / Action	Review Date	Add note
LR-001	OSHA	Warehouse	Yes	Sup	PPE	30/06/26	Ong
LR-002	Local Fire Law	Office	Yes	HR	Drill	15/06/26	Done
LR-003	ISO 45001	Maintenance	Yes	Safety	Inspection	20/06/26	Ong
______	______	______	______	______	______	______	______

3. Instructions for Use

- Assign **unique Requirement IDs** for each legal or regulatory obligation.

- Record **source of law, area/process, applicability, responsible person, control measures, review date, and status**.
- Use **short or single-word entries** to ensure compact table cells.
- Update the register whenever laws, regulations, or company operations change.
- Maintain as evidence for **ISO 45001 audits and internal reviews**.

Footer (appears on every page):
Page: ___ of ___ | **Confidential – Internal Use Only** | **Version:** ___ | **Year:** 20__

Certification Audit Readiness Evidence Index Template – Small Enterprises

Company: ________________ | **Document:** Certification Audit Evidence Index | **ISO 45001:2018**

Company Information

- Address: ________________________
- Email: _________________________
- Phone: ________________________

---

1. Purpose

This Evidence Index Template allows small enterprises to **track documents, records, and evidence** required for ISO 45001:2018 certification audits. It ensures all audit evidence is organized, accessible, and clearly linked to OH&S requirements, demonstrating compliance effectively.

---

2. Evidence Index Table

Evidence ID	Document / Record	ISO 45001 Clause	Location	Responsible	Available (Yes/No)	Reviewed (Yes/No)	Notes
EV-001	OH&S Policy	5.2	File	Manager	Yes	Yes	Updated
EV-002	Risk Assessment	6.1.2	File	Safety Officer	Yes	No	Review
EV-003	Training Matrix	7.2	File	HR	Yes	Yes	Current
EV-004	Incident Reports	10.2	File	Supervisor	Yes	No	Pending

EV-005	Internal Audit Report	9.2	File	Auditor	No	No	Planned
______	______	______	______	______	______	______	______

3. Instructions for Use

- Assign a **unique Evidence ID** for each document or record.
- Record the **document name, ISO 45001 clause, storage location, and responsible person**.
- Mark **Available** and **Reviewed** as Yes/No to track completeness and readiness.
- Use **Notes** for observations, pending updates, or version information.
- Keep updated to maintain **audit readiness** and demonstrate compliance during certification audits.
- Use **short or single-word entries** for compact table cells.

Footer (appears on every page):
Page: ___ of ___ | **Confidential – Internal Use Only** | **Version:** ___ | **Year:** 20__

Congratulations on completing the series of **OH&S templates and tools** designed for small enterprises. By using these fillable templates—including OH&S policies, risk assessments, training logs, nonconformity reports, corrective actions, root cause analysis, audit readiness checklists, and compliance registers—you now have a **structured, professional, and ISO 45001:2018-aligned system** to manage occupational health and safety effectively.

These templates are intended to:

- Simplify documentation and record-keeping
- Support proactive based on based on risk assessment methods defined by the organization in line with ISO 45001:2018 Clause 6.1.2" to align with risk-based thinking. and risk management
- Track employee training, competence, and awareness
- Monitor nonconformities and corrective actions
- Prepare your organization for internal and certification audits

Remember, an effective OH&S system is **more than documents**—it is about **consistent application, employee engagement, and continuous improvement**. Use these tools to empower your workforce, strengthen your safety culture, and maintain compliance with ISO 45001:2018.

Small steps taken consistently—documenting, tracking, verifying, and improving—will ensure a safer workplace, reduce incidents, and build confidence in your OH&S management system.

Kind regards,
Dr. Mohamed-Ali Ibrahim
Vienna, Austria

www.ingramcontent.com/pod-product-compliance
Lightning Source LLC
Chambersburg PA
CBHW041747220926
51847CB00013B/136
* 9 7 8 3 8 5 2 2 8 2 1 1 4 *